AF845951

Table of Contents

© Copyright 2023 - All rights reserved.

You may not reproduce, duplicate, or send the contents of this book without direct written permission from the author. You cannot hereby despite any circumstance blame the publisher or hold him or her to legal responsibility for any reparation, compensations, or monetary forfeiture owing to the information included herein, either in a direct or an indirect way.

Legal Notice: This book has copyright protection. You can use the book for personal purpose. You should not sell, use, alter, distribute, quote, take excerpts or paraphrase in part or whole the material contained in this book without obtaining the permission of the author first.

Disclaimer Notice: You must take note that the information in this document is for casual reading and entertainment purposes only. We have made every attempt to provide accurate, up to date and reliable information. We do not express or imply guarantees of any kind. The persons who read admit that the writer is not occupied in giving legal, financial, medical or other advice. We put this book content by sourcing various places.

Please consult a licensed professional before you try any techniques shown in this book. By going through this document, the book lover comes to an agreement that under no situation is the author accountable for any forfeiture, direct or indirect, which they may incur because of the use of material contained in this document, including, but not limited to, — errors, omissions, or inaccuracies.

PREFACE

I offer this book as a continuation of the memoirs of men of invention and industry published some years ago in the 'Lives of Engineers,' 'Industrial Biography,' and 'Self-Help.'

The early chapters relate to the history of a very important branch of British industry—that of Shipbuilding. A later chapter, kindly prepared by Sir Edward J. Harland, of Belfast, relates to the origin and progress of shipbuilding in Ireland.

Many of the facts set forth in the Life and Inventions of William Murdock have already been published in my 'Lives of Boulton and Watt;' but these are now placed in a continuous narrative, and supplemented by other information, more particularly the correspondence between Watt and Murdock, communicated to me by the present representative of the family, Mr. Murdock, C.E., of Gilwern, near Abergavenny.

I have also endeavoured to give as accurate an account as possible of the Invention of the Steam-printing Press, and its application to the production of Newspapers and Books,—an invention certainly of great importance to the spread of knowledge, science, and literature, throughout the world.

The chapter on the "Industry of Ireland" will speak for itself. It occurred to me, on passing through Ireland last year, that much remained to be said on that subject; and, looking to the increasing means of the country, and the well-known industry of its people, it seems reasonable to expect, that with peace, security, energy, and diligent labour of head and hand, there is really a great future before Ireland.

The last chapter, on "Astronomers in Humble Life," consists for the most part of a series of Autobiographies. It may seem, at first sight, to have little to do with the leading object of the book; but it serves to show what a number of active, earnest, and able men are comparatively hidden throughout society, ready to turn their hands and heads to the improvement of their own characters, if not to the advancement of the general community of which they form a part.

In conclusion, I say to the reader, as Quarles said in the preface to his 'Emblems,' "I wish thee as much pleasure in the reading as I had in the writing." In fact, the last three chapters were in some measure the cause of the book being published in its present form.

London, November, 1884.

CHAPTER I.

PHINEAS PETT: BEGINNINGS OF ENGLISH SHIP-BUILDING.

"A speck in the Northern Ocean, with a rocky coast, an ungenial climate, and a soil scarcely fruitful,—this was the material patrimony which descended to the English race—an inheritance that would have been little worth but for the inestimable moral gift that accompanied it. Yes; from Celts, Saxons, Danes, Normans—from some or all of them—have come down with English nationality a talisman that could command sunshine, and plenty, and empire, and fame. The 'go' which they transmitted to us—the national vis—this it is which made the old Angle-land a glorious heritage. Of this we have had a portion above our brethren—good measure, running over. Through this our island-mother has stretched out her arms till they enriched the globe of the earth....Britain, without her energy and enterprise, what would she be in Europe?"—Blackwood's Edinburgh Magazine (1870).

In one of the few records of Sir Isaac Newton's life which he left for the benefit of others, the following comprehensive thought occurs:

"It is certainly apparent that the inhabitants of this world are of a short date, seeing that all arts, as letters, ships, printing, the needle, &c., were discovered within the memory of history."

If this were true in Newton's time, how much truer is it now. Most of the inventions which are so greatly influencing, as well as advancing, the civilization of the world at the present time, have been discovered within the last hundred or hundred and fifty years. We do not say that man has become so much wiser during that period; for, though he has grown in Knowledge, the most fruitful of all things were said by "the heirs of all the ages" thousands of years ago.

But as regards Physical Science, the progress made during the last hundred years has been very great. Its most recent triumphs have been in connection with the discovery of electric power and electric light. Perhaps the most important invention, however, was that of the working steam engine, made by Watt only about a hundred years ago. The most recent application of this form of energy has been in the propulsion of ships, which has already produced so great an effect upon commerce, navigation, and the spread of population over the world.

Equally important has been the influence of the Railway—now the principal means of communication in all civilized countries. This invention has started into full life within our own time. The locomotive engine had for some years been employed in the haulage of coals; but it was not until the opening of the Liverpool and Manchester Railway in 1830, that the importance of the invention came to be acknowledged. The locomotive railway has since been everywhere adopted throughout Europe. In America, Canada, and the Colonies, it has opened up the boundless resources of the soil, bringing the country nearer to the towns, and the towns to the country. It has enhanced the celerity of time, and imparted a new series of conditions to every rank of life.

The importance of steam navigation has been still more recently ascertained. When it was first proposed, Sir Joseph Banks, President of the Royal Society, said: "It is a pretty plan, but there is just one point overlooked: that the steam-engine requires a firm basis on which to work." Symington, the practical mechanic, put this theory to the test by his successful experiments, first on Dalswinton Lake, and then on the Forth and Clyde Canal. Fulton and Bell afterwards showed the power of steamboats in navigating the rivers of America and Britain.

After various experiments, it was proposed to unite England and America by steam. Dr. Lardner, however, delivered a lecture before the Royal Institution in 1838, "proving" that steamers could never cross the Atlantic, because they could not carry sufficient coal to raise steam enough during the voyage. But this theory was also tested by experience in the same year, when the Sirius, of London, left Cork for New York, and made the passage in nineteen days. Four days after the departure of the Sirius, the Great Western left Bristol for New York, and made the passage in thirteen days five hours.[1] The problem was solved; and great ocean steamers have ever since passed in continuous streams between the shores of England and America.

In an age of progress, one invention merely paves the way for another. The first steamers were impelled by means of paddle wheels; but these are now almost entirely superseded by the screw. And this, too, is an invention almost of yesterday. It was only in 1840 that the Archimedes was fitted as a screw yacht.

A few years later, in 1845, the Great Britain, propelled by the screw, left Liverpool for New York, and made the voyage in fourteen days. The screw is now invariably adopted in all long ocean voyages.

It is curious to look back, and observe the small beginnings of maritime navigation. As regards this country, though its institutions are old, modern England is still young. As respects its mechanical and scientific achievements, it is the youngest of all countries. Watt's steam engine was the beginning of our manufacturing supremacy; and since its adoption, inventions and discoveries in Art and Science, within the last hundred years, have succeeded each other with extraordinary rapidity. In 1814 there was only one steam vessel in Scotland; while England possessed none at all. Now, the British mercantile steam-ships number about 5000, with about 4 millions of aggregate tonnage.[2]

In olden times this country possessed the materials for great things, as well as the men fitted to develope them into great results. But the nation was slow to awake and take advantage of its opportunities. There was no enterprise, no commerce—no "go" in the people. The roads were frightfully bad; and there was little communication between one part of the country and another.

If anything important had to be done, we used to send for foreigners to come and teach us how to do it. We sent for them to drain our fens, to build our piers and harbours, and even to pump our water at London Bridge. Though a seafaring population lived round our coasts, we did not fish our own seas, but left it to the industrious Dutchmen to catch the fish, and supply our markets. It was not until the year 1787 that the Yarmouth people began the deep-sea herring fishery; and yet these were the most enterprising amongst the English fishermen.

English commerce also had very slender beginnings. At the commencement of the fifteenth century, England was of very little account in the affairs of Europe. Indeed, the history of modern England is nearly coincident with the accession of the Tudors to the throne. With the exception of Calais and Dunkirk, her dominions on the Continent had been wrested from her by the French. The country at home had been made desolate by the Wars of the Roses. The population was very small, and had been kept down by war, pestilence, and famine.[3] The chief staple was wool, which was exported to Flanders in foreign ships, there to be manufactured into cloth. Nearly every article of importance was brought from abroad; and the little commerce which existed was in the hands of foreigners. The seas were swept by privateers, little better than pirates, who plundered without scruple every vessel, whether friend or foe, which fell in their way.

The British navy has risen from very low beginnings. The English fleet had fallen from its high estate since the reign of Edward III., who won a battle from the French and Flemings in 1340, with 260 ships; but his vessels were all of moderate size, being boats, yachts, and caravels, of very small tonnage. According to the contemporary chronicles, Weymouth, Fowey, Sandwich, and Bristol, were then of nearly almost as much importance as London;[4] which latter city only furnished twenty-five vessels, with 662 mariners.

The Royal Fleet began in the reign of Henry VII. Only six or seven vessels then belonged to the King, the largest being the Grace de Dieu, of comparatively small tonnage. The custom then was, to hire ships from the Venetians, the Genoese, the Hanse towns, and other trading people; and as soon as the service for which the vessels so hired was performed, they were dismissed.

When Henry VIII. ascended the throne in 1509, he directed his attention to the state of the navy. Although the insular position of England was calculated to stimulate the art of shipbuilding more than in most continental countries, our best ships long continued to be built by foreigners. Henry invited from abroad, especially from Italy, where the art of shipbuilding had made the greatest progress, as many skilful artists and workmen as he could procure, either by the hope of gain, or the high honours and distinguished countenance which he paid them. "By incorporating," says Charnock, "these useful persons among his own subjects, he soon formed a corps sufficient to rival those states which had rendered themselves most distinguished by their knowledge in this art; so that the fame of Genoa and Venice, which had long excited the envy of the greater part of Europe, became suddenly transferred to the shores of Britain."[5]

In fitting out his fleet, we find Henry disbursing large sums to foreigners for shipbuilding, for "harness" or armour, and for munitions of all sorts. The State Papers[6] particularize the amounts paid to Lewez de la Fava for "harness;" to William Gurre, "bregandy-maker;" and to Leonard Friscobald for "almayn ryvetts."

Francis de Errona, a Spaniard, supplied the gunpowder. Among the foreign mechanics and artizans employed were Hans Popenruyter, gunfounder of Mechlin; Robert Sakfeld, Robert Skorer, Fortuno de Catalenago, and John Cavelcant. On one occasion 2,797L. 19s. 4 1/2d. was disbursed for guns and grindstones. This sum must be multiplied by about four, to give the proper present value. Popenruyter seems to have been the great gunfounder of the age; he supplied the principal guns and gun stores for the English navy, and his name occurs in every Ordnance account of the series, generally for sums of the largest amounts.

Henry VIII. was the first to establish Royal dockyards, first at Woolwich, then at Portsmouth, and thirdly at Deptford, for the erection and repair of ships. Before then, England had been principally dependent upon Dutchmen and Venetians, both for ships of war and merchantmen. The sovereign had neither naval arsenals nor dockyards, nor any regular establishment of civil or naval affairs to provide ships of war. Sir Edward Howard, Lord High Admiral of England, at the accession of Henry VIII., actually entered into a "contract" with that monarch to fight his enemies.

This singular document is still preserved in the State Paper office. Even after the establishment of royal dockyards, the sovereign—as late as the reign of Elizabeth—entered into formal contracts with shipwrights for the repair and maintenance of ships, as well as for additions to the fleet.

The King, having made his first effort at establishing a royal navy, sent the fleet to sea against the ships of France. The Regent was the ship royal, with Sir Thomas Knivet, Master of the Horse, and Sir John Crew of Devonshire, as Captains. The fleet amounted to twenty-five well furnished ships. The French fleet were thirty-nine in number. They met in Brittany Bay, and had a fierce fight. The Regent grappled with a great carack of Brest; the French, on the English boarding their ship, set fire to the gunpowder, and both ships were blown up, with all their men. The French fleet fled, and the English kept the seas. The King, hearing of the loss of the Regent, caused a great ship to be built, the like of which had never before been seen in England, and called it Harry Grace de Dieu.

This ship was constructed by foreign artizans, principally by Italians, and was launched in 1515. She was said to be of a thousand tons portage—the largest ship in England. The vessel was four-masted, with two round tops on each mast, except the shortest mizen. She had a high forecastle and poop, from which the crew could shoot down upon the deck or waist of another vessel. The object was to have a sort of castle at each end of the ship. This style of shipbuilding was doubtless borrowed from the Venetians, then the greatest naval power in Europe. The length of the masts, the height of the ship above the water's edge, and the ornaments and decorations, were better adapted for the stillness of the Adriatic and Mediterranean Seas, than for the boisterous ocean of the northern parts of Europe.[7] The story long prevailed that "the Great Harry swept a dozen flocks of sheep off the Isle of Man with her bob-stay." An American gentleman (N.B. Anderson, LL.D., Boston) informed the present author that this saying is still proverbial amongst the United States sailors.

The same features were reproduced in merchant ships. Most of them were suited for defence, to prevent the attacks of pirates, which swarmed the seas round the coast at that time. Shipbuilding by the natives in private shipyards was in a miserable condition. Mr. Willet, in his memoir relative to the navy, observes: "It is said, and I believe with truth, that at this time (the middle of the sixteenth century) there was not a private builder between London Bridge and Gravesend, who could lay down a ship in the mould left from a Navy Board's draught, without applying to a tinker who lived in Knave's Acre."[8]

Another ship of some note built at the instance of Henry VIII. was the Mary Rose, of the portage of 500 tons. We find her in the "pond at Deptford" in 1515. Seven years later, in the thirtieth year of Henry VIII.'s reign, she was sent to sea, with five other English ships of war, to protect such commerce as then existed from the depredations of the French and Scotch pirates. The Mary Rose was sent many years later (in 1544) with the English fleet to the coast of France, but returned with the rest of the fleet to Portsmouth without entering into any engagement. While laid at anchor, not far from the place where the Royal George afterwards went down, and the ship was under repair, her gun-ports being very low when she was laid over, "the shipp turned, the water entered, and sodainly she sanke."

What was to be done? There were no English engineers or workmen who could raise the ship. Accordingly, Henry VIII. sent to Venice for assistance, and when the men arrived, Pietro de Andreas was dispatched with the Venetian marines and carpenters to raise the Mary Rose. Sixty English mariners were appointed to attend upon them. The Venetians were then the skilled "heads," the English were only the "hands." Nevertheless they failed with all their efforts; and it was not until the year 1836 that Mr. Dean, the engineer, succeeded in raising not only the Royal George, but the Mary Rose, and cleared the roadstead at Portsmouth of the remains of the sunken ships.

When Elizabeth ascended the throne in 1558, the commerce and navigation of England were still of very small amount. The population of the kingdom amounted to only about five millions—not much more than the population of London is now. The country had little commerce, and what it had was still mostly in the hands of foreigners. The Hanse towns had their large entrepot for merchandise in Cannon Street, on the site of the present Cannon Street Station. The wool was still sent abroad to Flanders to be fashioned into cloth, and even garden produce was principally imported from Holland. Dutch, Germans, Flemings, French, and Venetians continued to be our principal workmen. Our iron was mostly obtained from Spain and Germany. The best arms and armour came from France and Italy. Linen was imported from Flanders and Holland, though the best came from Rheims. Even the coarsest dowlas, or sailcloth, was imported from the Low Countries.

The royal ships continued to be of very small burthen, and the mercantile ships were still smaller. The Queen, however, did what she could to improve the number and burthen of our ships. "Foreigners," says Camden, "stiled her the restorer of naval glory and Queen of the Northern Seas." In imitation of the Queen, opulent subjects built ships of force; and in course of time England no longer depended upon Hamburg, Dantzic, Genoa, and Venice, for her fleet in time of war.

Spain was then the most potent power in Europe, and the Netherlands, which formed part of the dominions of Spain, was the centre of commercial prosperity. Holland possessed above 800 good ships, of from 200 to 700 tons burthen, and above 600 busses for fishing, of from 100 to 200 tons. Amsterdam and Antwerp were in the heyday of their prosperity. Sometimes 500 great ships were to be seen lying together before Amsterdam;[9] whereas England at that time had not four merchant ships of 400 tons each! Antwerp, however, was the most important city in the Low Countries. It was no uncommon thing to see as many as 2500 ships in the Scheldt, laden with merchandize. Sometimes 500 ships would come and go from Antwerp in one day, bound to or returning from the distant parts of the world. The place was immensely rich, and was frequented by Spaniards, Germans, Danes, English, Italians, and Portuguese the Spaniards being the most numerous. Camden, in his history of Queen Elizabeth, relates that our general trade with the Netherlands in 1564 amounted to twelve millions of ducats, five millions of which was for English cloth alone.

The religious persecutions of Philip II. of Spain and of Charles IX. of France shortly supplied England with the population of which she stood in need—active, industrious, intelligent artizans. Philip set up the Inquisition in Flanders, and in a few years more than 50,000 persons were deliberately murdered. The Duchess of Parma, writing to Philip II. in 1567, informed him that in a few days above 100,000 men had already left the country with their money and goods, and that more were following every day. They fled to Germany, to Holland, and above all to England, which they hailed as Asylum Christi. The emigrants settled in the decayed cities and towns of Canterbury, Norwich, Sandwich, Colchester, Maidstone, Southampton, and many other places, where they carried on their manufactures of woollen, linen, and silk, and established many new branches of industry.[10]

Five years later, in 1572, the massacre of St. Bartholomew took place in France, during which the Roman Catholic Bishop Perefixe alleges that 100,000 persons were put to death because of their religions opinions. All this persecution, carried on so near the English shores, rapidly increased the number of foreign fugitives into England, which was followed by the rapid advancement of the industrial arts in this country.

The asylum which Queen Elizabeth gave to the persecuted foreigners brought down upon her the hatred of Philip II. and Charles IX. When they found that they could not prevent her furnishing them with an asylum, they proceeded to compass her death. She was excommunicated by the Pope, and Vitelli was hired to assassinate her. Philip also proceeded to prepare the Sacred Armada for the subjugation of the English nation, and he was master of the most powerful army and navy in the world.

Modern England was then in the throes of her birth. She had not yet reached the vigour of her youth, though she was full of life and energy. She was about to become the England of free thought, commerce, and manufactures; to plough the ocean with her navies, and to plant her colonies over the earth. Up to the accession of Elizabeth, she had done little, but now she was about to do much.

It was a period of sudden emancipation of thought, and of immense fertility and originality. The poets and prose writers of the time united the freshness of youth with the vigour of manhood. Among these were Spenser, Shakespeare, Sir Philip Sidney, the Fletchers, Marlowe, and Ben Jonson. Among the statesmen of Elizabeth were Burleigh, Leicester, Walsingham, Howard, and Sir Nicholas Bacon. But perhaps greatest of all were the sailors, who, as Clarendon said, "were a nation by themselves;" and their leaders—Drake, Frobisher, Cavendish, Hawkins, Howard, Raleigh, Davis, and many more distinguished seamen.

They were the representative men of their time, the creation in a great measure of the national spirit. They were the offspring of long generations of seamen and lovers of the sea. They could not have been great but for the nation which gave them birth, and imbued them with their worth and spirit. The great sailors, for instance, could not have originated in a nation of mere landsmen.

They simply took the lead in a country whose coasts were fringed with sailors. Their greatness was but the result of an excellence in seamanship which prevailed widely around them.

The age of English maritime adventure only began in the reign of Elizabeth. England had then no colonies—no foreign possessions whatever. The first of her extensive colonial possessions was established in this reign. "Ships, colonies, and commerce" began to be the national motto—not that colonies make ships and commerce, but that ships and commerce make colonies. Yet what cockle-shells of ships our pioneer navigators first sailed in!

Although John Cabot or Gabota, of Bristol, originally a citizen of Venice, had discovered the continent of North America in 1496, in the reign of Henry VII., he made no settlement there, but returned to Bristol with his four small ships. Columbus did not see the continent of America until two years later, in 1498, his first discoveries being the islands of the West Indies.

It was not until the year 1553 that an attempt was made to discover a North-west passage to Cathaya or China. Sir Hugh Willonghby was put in command of the expedition, which consisted of three ships,—the Bona Esperanza, the Bona Ventura (Captain Chancellor), and the Bona

Confidentia (Captain Durforth),—most probably ships built by Venetians. Sir Hugh reached 72 degrees of north latitude, and was compelled by the buffeting of the winds to take refuge with Captain Durforth's vessel at Arcina Keca, in Russian Lapland, where the two captains and the crews of these ships, seventy in number, were frozen to death. In the following year some Russian fishermen found Sir John Willonghby sitting dead in his cabin, with his diary and other papers beside him.

Captain Chancellor was more fortunate. He reached Archangel in the White Sea, where no ship had ever been seen before. He pointed out to the English the way to the whale fishery at Spitzbergen, and opened up a trade with the northern parts of Russia. Two years later, in 1556, Stephen Burroughs sailed with one small ship, which entered the Kara Sea; but he was compelled by frost and ice to return to England. The strait which he entered is still called "Burrough's Strait."

It was not, however, until the reign of Elizabeth that great maritime adventures began to be made. Navigators were not so venturous as they afterwards became. Without proper methods of navigation, they were apt to be carried away to the south, across an ocean without limit. In 1565 a young captain, Martin Frobisher, came into notice. At the age of twenty-five he captured in the South Seas the Flying Spirit, a Spanish ship laden with a rich cargo of cochineal. Four years later, in 1569, he made his first attempt to discover the north-west passage to the Indies, being assisted by Ambrose Dudley, Earl of Warwick. The ships of Frobisher were three in number, the Gabriel, of from 15 to 20 tons; the Michael, of from 20 to 25 tons, or half the size of a modern fishing-boat; and a pinnace, of from 7 to 10 tons! The aggregate of the crews of the three ships was only thirty-five, men and boys. Think of the daring of these early navigators in attempting to pass by the North Pole to Cathay through snow, and storm, and ice, in such miserable little cockboats! The pinnace was lost; the Michael, under Owen Griffith, a Welsh-man, deserted; and Martin Frobisher in the Gabriel went alone into the north-western sea!

He entered the great bay, since called Hudson's Bay, by Frobisher's Strait. He returned to England without making the discovery of the Passage, which long remained the problem of arctic voyagers. Yet ten years later, in 1577, he made another voyage, and though he made his second attempt with one of Queen Elizabeth's own ships, and two barks, with 140 persons in all, he was as unsuccessful as before. He brought home some supposed gold ore; and on the strength of the stones containing gold, a third expedition went out in the following year. After losing one of the ships, consuming the provisions, and suffering greatly from ice and storms, the fleet returned home one by one. The supposed gold ore proved to be only glittering sand.

While Frobisher was seeking El-Dorado in the North, Francis Drake was finding it in the South. He was a sailor, every inch of him.

"Pains, with patience in his youth," says Fuller, "knit the joints of his soul, and made them more solid and compact." At an early age, when carrying on a coasting trade, his imagination was inflamed by the exploits of his protector Hawkins in the New World, and he joined him in his last unfortunate adventure on the Spanish Main. He was not, however, discouraged by his first misfortune, but having assembled about him a number of seamen who believed in him, he made other adventures to the West Indies, and learnt the navigation of that part of the ocean. In 1570, he obtained a regular commission from Queen Elizabeth, though he sailed his own ships, and made his own ventures. Every Englishman, who had the means, was at liberty to fit out his own ships; and with tolerable vouchers, he was able to procure a commission from the Court, and proceed to sea at his own risk and cost. Thus, the naval enterprise and pioneering of new countries under Elizabeth, was almost altogether a matter of private enterprise and adventure.

In 1572, the butchery of the Hugnenots took place at Paris and throughout France; while at the same time the murderous power of Philip II. reigned supreme in the Netherlands. The sailors knew what they had to expect from the Spanish king in the event of his obtaining his threatened revenge upon England; and under their chosen chiefs they proceeded to make war upon him. In the year of the massacre of St. Bartholomew, Drake set sail for the Spanish Main in the Pasha, of seventy tons, accompanied by the Swan, of twenty-five tons; the united crews of the vessels amounting to seventy-three men and boys. With this insignificant force, Drake made great havoc amongst the Spanish shipping at Nombre de Dios. He partially crossed the Isthmus of Darien, and obtained his first sight of the great Pacific Ocean. He returned to England in August 1573, with his frail barks crammed with treasure.

A few years later, in 1577, he made his ever-memorable expedition. Charnock says it was "an attempt in its nature so bold and unprecedented, that we should scarcely know whether to applaud it as a brave, or condemn it as a rash one, but for its success." The squadron with which he sailed for South America consisted of five vessels, the largest of which, the Pelican, was only of 100 tons burthen; the next, the Elizabeth, was of 80; the third, the Swan, a fly-boat, was of 50; the Marygold bark, of 30; and the Christopher, a pinnace, of 15 tons. The united crews of these vessels amounted to only 164, gentlemen and sailors.

The gentlemen went with Drake "to learn the art of navigation." After various adventures along the South American coast, the little fleet passed through the Straits of Magellan, and entered the Pacific Ocean. Drake took an immense amount of booty from the Spanish towns along the coast, and captured the royal galleon, the Cacafuego, laden with treasure. After trying in vain to discover a passage home by the North-eastern ocean, though what is now known as Behring Straits, he took shelter in Port San Francisco, which he took possession of in the name of the Queen of England, and called New Albion. He eventually crossed the Pacific for the Moluccas and Java, from which he sailed right across the Indian Ocean, and by the Cape of Good Hope to England, thus making the circumnavigation of the world. He was absent with his little fleet for about two years and ten months.

Not less extraordinary was the voyage of Captain Cavendish, who made the circumnavigation of the globe at his own expense. He set out from Plymouth in three small vessels on the 21st July, 1586. One vessel was of 120 tons, the second of 60 tons, and the third of 40 tons—not much bigger than a Thames yacht. The united crews, of officers, men, and boys, did not exceed 123! Cavendish sailed along the South American continent, and made through the Straits of Magellan, reaching the Pacific Ocean. He burnt and plundered the Spanish settlements along the coast, captured some Spanish ships, and took by boarding the galleon St. Anna, with 122,000 Spanish dollars on board. He then sailed across the Pacific to the Ladrone Islands, and returned home through the Straits of Java and the Indian Archipelago by the Cape of Good Hope, and reached England after an absence of two years and a month.

The sacred and invincible Armada was now ready, Philip II. was determined to put down those English adventurers who had swept the coasts of Spain and plundered his galleons on the high seas. The English sailors knew that the sword of Philip was forged in the gold mines of South America, and that the only way to defend their country was to intercept the plunder on its voyage home to Spain. But the sailors and their captains—Drake, Hawkins, Frobisher, Howard, Grenville, Raleigh, and the rest—could not altogether interrupt the enterprise of the King of Spain. The Armada sailed, and came in sight of the English coast on the 20th of July, 1588.

The struggle was of an extraordinary character. On the one side was the most powerful naval armament that had ever put to sea. It consisted of six squadrons of sixty fine large ships, the smallest being of 700 tons. Besides these were four gigantic galleasses, each carrying fifty guns, four large armed galleys, fifty-six armed merchant ships, and twenty caravels—in all, 149 vessels. On board were 8000 sailors, 20,000 soldiers, and a large number of galley-slaves. The ships carried provisions enough for six months' consumption; and the supply of ammunition was enormous.

On the other side was the small English fleet under Hawkins and Drake. The Royal ships were only thirteen in number. The rest were contributed by private enterprize, there being only thirty-eight vessels of all sorts and sizes, including cutters and pinnaces, carrying the Queen's flag. The principal armed merchant ships were provided by London, Southampton, Bristol, and the other southern ports. Drake was followed by some privateers; Hawkins had four or five ships, and Howard of Effingham two. The fleet was, however, very badly found in provisions and ammunition. There was only a week's provisions on board, and scarcely enough ammunition for one day's hard fighting. But the ships, small though they were, were in good condition. They could sail, whether in pursuit or in flight, for the men who navigated them were thorough sailors.

The success of the defence was due to tact, courage, and seamanship. At the first contact of the fleets, the Spanish towering galleons wished to close, to grapple with their contemptuous enemies, and crush them to death. "Come on!" said Medina Sidonia. Lord Howard came on with the Ark and three other ships, and fired with immense rapidity into the great floating castles. The Sam Mateo luffed, and wanted them to board. "No! not yet!" The English tacked, returned, fired again, riddled the Spaniards, and shot away in the eye of the wind. To the astonishment of the Spanish Admiral, the English ships approached him or left him just as they chose. "The enemy pursue me," wrote the Spanish Admiral to the Prince of Parma; "they fire upon me most days from morning till nightfall, but they will not close and grapple, though I have given them every opportunity." The Capitana, a galleon of 1200 tons, dropped behind, struck her flag to Drake, and increased the store of the English fleet by some tons of gunpowder. Another Spanish ship surrendered, and another store of powder and shot was rescued for the destruction of the Armada. And so it happened throughout, until the Spanish fleet was driven to wreck and ruin, and the remaining ships were scattered by the tempests of the north. After all, Philip proved to be, what the sailors called him, only "a Colossus stuffed with clouts."

The English sailors followed up their advantage. They went on "singeing the Ring of Spain's beard." Private adventurers fitted up a fleet under the command of Drake, and invaded the mainland of Spain. They took the lower part of the town of Corunna; sailed to the Tagus, and captured a fleet of ships laden with wheat and warlike stores for a new Armada. They next sacked Vigo, and returned to England with 150 pieces of cannon and a rich booty. The Earl of Cumberland sailed to the West Indies on a private adventure, and captured more Spanish prizes. In 1590, ten English merchantmen, returning from the Levant, attacked twelve Spanish galleons, and after six hours' contest, put them to flight with great loss. In the following year, three merchant ships set sail for the East Indies, and in the course of their voyage took several Portuguese vessels.

A powerful Spanish fleet still kept the seas, and in 1591 they conquered the noble Sir Richard Grenville at the Azores—fifteen great Spanish galleons against one Queen's ship, the Revenge. In 1593, two of the Queen's ships, accompanied by a number of merchant ships, sailed for the West Indies, under Burroughs, Frobisher, and Cross, and amongst their other captures they took the greatest of all the East India caracks, a vessel of 1600 tons, 700 men, and 36 brass cannon, laden with a magnificent cargo. She was taken to Dartmouth, and surprised all who saw her, being the largest ship that had ever been seen in England. In 1594, Captain James Lancaster set sail with three ships upon a voyage of adventure. He was joined by some Dutch and French privateers. The result was, that they captured thirty-nine of the Spanish ships. Sir Amias Preston, Sir John Hawkins, and Sir Francis Drake, also continued their action upon the seas. Lord Admiral Howard and the Earl of Essex made their famous attack upon Cadiz for the purpose of destroying the new Armada; they demolished all the forts; sank eleven of the King of Spain's best ships, forty-four merchant ships, and brought home much booty.

Nor was maritime discovery neglected. The planting of new colonies began, for the English people had already begun to swarm. In 1578, Sir Humphrey Gilbert planted Newfoundland for the Queen. In 1584, Sir Waiter Raleigh planted the first settlement in Virginia. Nor was the North-west passage neglected; for in 1580, Captain Pett (a name famous on the Thames) set sail from Harwich in the George, accompanied by Captain Jackman in the William. They reached the ice in the North Sea, but were compelled to return without effecting their purpose! Will it be believed that the George was only of 40 tons, and that its crew consisted of nine men and a boy; and that the William was of 20 tons, with five men and a boy? The wonder is that these little vessels could resist the terrible icefields, and return to England again with their hardy crews.

Then in 1585, another of our adventurous sailors, John Davis, of Sandridge on the Dart, set sail with two barks, the Sunshine and the Moonshine, of 50 and 35 tons respectively, and discovered in the far North-west the Strait which now bears his name. He was driven back by the ice; but, undeterred by his failure, he set out on a second, and then on a third voyage of discovery in the two following years. But he never succeeded in discovering the North-west passage. It all reads like a mystery—these repeated, determined, and energetic attempts to discover a new way of reaching the fabled region of Cathay.

In these early times the Dutch were not unworthy rivals of the English. After they had succeeded in throwing off the Spanish yoke and achieved their independence, they became one of the most formidable of maritime powers. In the course of another century Holland possessed more colonies, and had a larger share of the carrying trade of the world than Britain. It was natural therefore that the Dutch republic should take an interest in the North-west passage; and the Dutch sailors, by their enterprise and bravery, were among the first to point the way to Arctic discovery. Barents and Behring, above all others, proved the courage and determination of their heroic ancestors.

The romance of the East India Company begins with an advertisement in the London Gazette of 1599, towards the end of the reign of Queen Elizabeth. As with all other enterprises of the nation, it was established by private means. The Company was started with a capital of 72,000L. in 50L. shares. The adventurers bought four vessels of an average burthen of 350 tons. These were stocked with provisions, "Norwich stuffs," and other merchandise. The tiny fleet sailed from Billingsgate on the 13th February, 1601. It went by the Cape of Good Hope to the East Indies, under the command of Captain James Lancaster. It took no less than sixteen months to reach the Indian Archipelago.

The little fleet reached Acheen in June, 1602. The king of the territory received the visitors with courtesy, and exchanged spices with them freely. The four vessels sailed homeward, taking possession of the island of St. Helena on their way back; having been absent exactly thirty-one months. The profits of the first voyage proved to be about one hundred per cent. Such was the origin of the great East India Company—now expanded into an empire, and containing about two hundred millions of people.

To return to the shipping and the mercantile marine of the time of Queen Elizabeth. The number of Royal ships was only thirteen, the rest of the navy consisting of merchant ships, which were hired and discharged when their purpose was served.[11] According to Wheeler, at the accession of the Queen, there were not more than four ships belonging to the river Thames, excepting those of the Royal Navy, which were over 120 tons in burthen;[12] and after forty years, the whole of the merchant ships of England, over 100 tons, amounted to 135; only a few of these being of 500 tons. In 1588, the number had increased to 150, "of about 150 tons one with another, employed in trading voyages to all parts and countries." The principal shipping which frequented the English ports still continued to be foreign—Italian, Flemish, and German.

Liverpool, now possessing the largest shipping tonnage in the world, had not yet come into existence. It was little better than a fishing village. The people of the place presented a petition to the Queen, praying her to remit a subsidy which had been imposed upon them, and speaking of their native place as "Her Majesty's poor decayed town of Liverpool." In 1565, seven years after Queen Elizabeth began to reign, the number of vessels belonging to Liverpool was only twelve. The largest was of forty tons burthen, with twelve men; and the smallest was a boat of six tons, with three men.[13]

James I., on his accession to the throne of England in 1603, called in all the ships of war, as well as the numerous privateers which had been employed during the previous reign in waging war against the commerce of Spain, and declared himself to be at peace with all the world. James was as peaceful as a Quaker. He was not a fighting King;—and, partly on this account, he was not popular. He encouraged manufactures in wool, silk, and tapestry. He gave every encouragement to the mercantile and colonizing adventurers to plant and improve the rising settlements of Virginia, New England, and Newfoundland. He also promoted the trade to the East Indies. Attempts continued to be made, by Hudson, Poole, Button, Hall, Baffin, and other courageous seamen, to discover the North-West passage, but always without effect.

The shores of England being still much infested by Algerine and other pirates,[14] King James found it necessary to maintain the ships of war in order to protect navigation and commerce. He nearly doubled the ships of the Royal Navy, and increased the number from thirteen to twenty-four. Their size, however, continued small, both Royal and merchant ships. Sir William Monson says, that at the accession of James I. there were not above four merchant ships in England of 400 tons burthen.[15] The East Indian merchants were the first to increase the size. In 1609, encouraged by their Charter, they built the Trade's Increase, of 1100 tons burthen, the largest merchant ship that had ever been built in England. As it was necessary that, the crew of the ship should be able to beat off the pirates, she was fully armed. The additional ships of war were also of heavier burthen. In the same year, the Prince, of 1400 tons burthen, was launched; she carried sixty-four cannon, and was superior to any ship of the kind hitherto seen in England.

And now we arrive at the subject of this memoir. The Petts were the principal ship-builders of the time. They had long been known upon the Thames, and had held posts in the Royal Dockyards since the reign of Henry VII. They were gallant sailors, too; one of them, as already mentioned, having made an adventurous voyage to the Arctic Ocean in his little bark, the George, of only 40 tons burthen. Phineas Pett was the first of the great ship-builders. His father, Peter Pett, was one of the Queen's master shipwrights. Besides being a ship-builder, he was also a poet, being the author of a poetical piece entitled, "Time's Journey to seek his daughter Truth,"[16] a very respectable performance. Indeed, poetry is by no means incompatible with ship-building—the late Chief Constructor of the Navy being, perhaps, as proud of his poetry as of his ships. Pett's poem was dedicated to the Lord High Admiral, Howard, Earl of Nottingham; and this may possibly have been the reason of the singular interest which he afterwards took in Phineas Pett, the poet shipwright's son.

Phineas Pett was the second son of his father. He was born at Deptford, or "Deptford Strond," as the place used to be called, on the 1st of November, 1570. At nine years old, he was sent to the free-school at Rochester, and remained there for four years. Not profiting much by his education there, his father removed him to a private school at Greenwich, kept by a Mr. Adams. Here he made so much progress, that in three years time he was ready for Cambridge. He was accordingly sent to that University at Shrovetide, 1586, and was entered at Emmanuel College, under charge of Mr. Charles Chadwick, the president. His father allowed him 20L. per annum, besides books, apparel, and other necessaries.

Phineas remained at Cambridge for three years. He was obliged to quit the University by the death of his "reverend, ever-loving father," whose loss, he says, "proved afterwards my utter undoing almost, had not God been more merciful to me." His mother married again, "a most wicked husband," says Pett in his autobiography,[17] "one, Mr. Thomas Nunn, a minister," but of what denomination he does not state. His mother's imprudence wholly deprived him of his maintenance, and having no hopes of preferment from his friends, he necessarily abandoned his University career, "presently after Christmas, 1590."

Early in the following year, he was persuaded by his mother to apprentice himself to Mr. Richard Chapman, of Deptford Strond, one of the Queen's Master shipwrights, whom his late father had "bred up from a child to that profession." He was allowed 2L. 6s. 8d. per annum, with which he had to provide himself with tools and apparel. Pett spent two years in this man's service to very little purpose; Chapman then died, and the apprentice was dismissed. Pett applied to his elder brother Joseph, who would not help him, although he had succeeded to his father's post in the Royal Dockyard. He was accordingly "constrained to ship himself to sea upon a desperate voyage in a man-of-war." He accepted the humble place of carpenter's mate on board the galleon Constance, of London. Pett's younger brother, Peter, then living at Wapping, gave him lodging, meat, and drink, until the ship was ready to sail. But he had no money to buy clothes. Fortunately one William King, a yoeman in Essex, taking pity upon the unfortunate young man, lent him 3L. for that purpose; which Pett afterwards repaid.

The Constance was of only 200 tons burden. She set sail for the South a few days before Christmas, 1592. There is no doubt that she was bound upon a piratical adventure. Piracy was not thought dishonourable in those days. Four years had elapsed since the Armada had approached the English coast; and now the English and Dutch ships were scouring the seas in search of Spanish galleons.

Whoever had the means of furnishing a ship, and could find a plucky captain to command her, sent her out as a privateer. Even the Companies of the City of London clubbed their means together for the purpose of sending out Sir Waiter Raleigh to capture Spanish ships, and afterwards to divide the plunder; as any one may see on referring to the documents of the London Corporation.[18]

The adventure in which Pett was concerned did not prove very fortunate. He was absent for about twenty months on the coasts of Spain and Barbary, and in the Levant, enduring much misery for want of victuals and apparel, and "without taking any purchase of any value." The

Constance returned to the Irish coast, "extreme poorly." The vessel entered Cork harbour, and then Pett, thoroughly disgusted with privateering life, took leave of both ship and voyage. With much difficulty, he made his way across the country to Waterford, from whence he took ship for London. He arrived there three days before Christmas, 1594, in a beggarly condition, and made his way to his brother Peter's house at Wapping, who again kindly entertained him. The elder brother Joseph received him more coldly, though he lent him forty shillings to find himself in clothes. At that time, the fleet was ordered to be got ready for the last expedition of Drake and Hawkins to the West Indies. The Defiance was sent into Woolwich dock to be sheathed; and as Joseph Pett was in charge of the job, he allowed his brother to be employed as a carpenter.

In the following year, Phineas succeeded in attracting the notice of Matthew Baker, who was commissioned to rebuild Her Majesty's Triumph. Baker employed Pett as an ordinary workman; but he had scarcely begun the job before Baker was ordered to proceed with the building of a great new ship at Deptford, called the Repulse.

Phineas wished to follow the progress of the Triumph, but finding his brother Joseph unwilling to retain him in his employment, he followed Baker to Deptford, and continued to work at the Repulse until she was finished, launched, and set sail on her voyage, at the end of April, 1596. This was the leading ship of the squadron which set sail for Cadiz, under the command of the Earl of Essex and the Lord Admiral Howard, and which did so much damage to the forts and shipping of Philip II. of Spain.

During the winter months, while the work was in progress, Pett spent the leisure of his evenings in perfecting himself in learning, especially in drawing, cyphering, and mathematics, for the purpose, as he says, of attaining the knowledge of his profession. His master, Mr. Baker, gave him every encouragement, and from his assistance, he adds, "I must acknowledge I received my greatest lights." The Lord Admiral was often present at Baker's house. Pett was importuned to set sail with the ship when finished, but he preferred remaining at home. The principal reason, no doubt, that restrained him at this moment from seeking the patronage of the great, was the care of his two sisters,[19] who, having fled from the house of their barbarous stepfather, could find no refuge but in that of their brother Phineas. Joseph refused to receive them, and Peter of Wapping was perhaps less able than willing to do so.

In April, 1597, Pett had the advantage of being introduced to Howard, Earl of Nottingham, then Lord High Admiral of England. This, he says, was the first beginning of his rising. Two years later, Howard recommended him for employment in purveying plank and timber in Norfolk and Suffolk for shipbuilding purposes. Pett accomplished his business satisfactorily, though he had some malicious enemies to contend against. In his leisure, he began to prepare models of ships, which he rigged and finished complete. He also proceeded with the study of mathematics. The beginning of the year 1600 found Pett once more out of employment; and during his enforced idleness, which continued for six months, he seriously contemplated abandoning his profession and attempting to gain "an honest and convenient maintenance" by joining a friend in purchasing a caravel (a small vessel), and navigating it himself.

He was, however, prevented from undertaking this enterprise by a message which he received from the Court, then stationed at Greenwich. The Lord High Admiral desired to see him; and after many civil compliments, he offered him the post of keeper of the plankyard at Chatham. Pett was only too glad to accept this offer, though the salary was small. He shipped his furniture on board a hoy of Rainham, and accompanied it down the Thames to the junction with the Medway. There he escaped a great danger—one of the sea perils of the time. The mouths of navigable rivers were still infested with pirates; and as the hoy containing Pett approached the Nore about three o'clock in the morning, and while still dark, she came upon a Dunkirk picaroon, full of men. Fortunately the pirate was at anchor; she weighed and gave chase, and had not the hoy set full sail, and been impelled up the Swale by a fresh wind, Pett would have been taken prisoner, with all his furniture.[20]

Arrived at Chatham, Pett met his brother Joseph, became reconciled to him, and ever after they lived together as loving brethren. At his brother's suggestion, Pett took a lease of the Manor House, and settled there with his sisters. He was now in the direct way to preferment. Early in the following year (March, 1601) he succeeded to the place of assistant to the principal master shipwright at Chatham, and undertook the repairs of Her Majesty's ship The Lion's Whelp, and in the next year he new-built the Moon enlarging her both in length and breadth.

At the accession of James I. in 1603, Pett was commanded by the Lord High Admiral with all possible speed to build a little vessel for the young Prince Henry, eldest son of His Majesty. It was to be a sort of copy of the Ark Royal, which was the flagship of the Lord High Admiral when he defeated the Spanish Armada. Pett proceeded to accomplish the order with all dispatch. The little ship was in length by the keel 28 feet, in breadth 12 feet, and very curiously garnished within and without with painting and carving. After working by torch and candle light, night and day, the ship was launched, and set sail for the Thames, with the noise of drums, trumpets, and cannon, at the beginning of March, 1604. After passing through a great storm at the Nore, the vessel reached the Tower, where the King and the young Prince inspected her with delight. She was christened Disdain by the Lord High Admiral, and Pett was appointed captain of the ship.

After his return to Chatham, Pett, at his own charge, built a small ship at Gillingham, of 300 tons, which he launched in the same year, and named the Resistance. The ship was scarcely out of hand, when Pett was ordered to Woolwich, to prepare the Bear and other vessels for conveying his patron, the Lord High Admiral, as an Ambassador Extraordinary to Spain, for the purpose of concluding peace, after a strife of more than forty years. The Resistance was hired by the Government as a transport, and Pett was put in command. He seems to have been married at this time, as he mentions in his memoir that he parted with his wife and children at Chatham on the 24th of March, 1605, and that he sailed from Queenborough on Easter Sunday.

During the voyage to Lisbon the Resistance became separated from the Ambassador's squadron, and took refuge in Corunna. She then set sail for Lisbon, which she reached on the 24th of April; and afterwards for St. Lucar, on the Guadalquiver, near Seville, which she reached on the 11th of May following. After revisiting Corunna, "according to instructions," on the homeward voyage, Pett directed his course for England, and reached Rye on the 26th of June, "amidst much rain, thunder, and lightning." In the course of the same year, his brother Joseph died, and Phineas succeeded to his post as master shipbuilder at Chatham. He was permitted, in conjunction with one Henry Farvey and three others, to receive the usual reward of 5s. per ton for building five new merchant ships,[21] most probably for East Indian commerce, now assuming large dimensions. He was despatched by the Government to Bearwood, in Hampshire, to make a selection of timber from the estate of the Earl of Worcester for the use of the navy, and on presenting his report 3000 tons were purchased. What with his building of ships, his attendance on the Lord Admiral to Spain, and his selection of timber for the Government, his hands seem to have been kept very full during the whole of 1605.

In July, 1606, Pett received private instructions from the Lord High Admiral to have all the King's ships "put into comely readiness" for the reception of the King of Denmark, who was expected on a Royal visit. "Wherein," he says, "I strove extraordinarily to express my service for the honour of the kingdom; but by reason the time limited was short, and the business great, we laboured night and day to effect it, which accordingly was done, to the great honour of our sovereign king and master, and no less admiration of all strangers that were eye-witnesses to the same." The reception took place on the 10th of August, 1606.

Shortly after the departure of His Majesty of Denmark, four of the Royal ships—the Ark, Victory, Golden Lion, and Swiftsure—were ordered to be dry-docked; the two last mentioned at Deptford, under charge of Matthew Baker; and the two former at Woolwich, under that of Pett. For greater convenience, Pett removed his family to Woolwich. After being elected and sworn Master of the Company of Shipwrights, he refers in his manuscript, for the first time, to his magnificent and original design of the Prince Royal.[22]

"After settling at Woolwich," he says, "I began a curious model for the prince my master, most part whereof I wrought with my own hands." After finishing the model, he exhibited it to the Lord High Admiral, and, after receiving his approval and commands, he presented it to the young prince at Richmond. "His Majesty (who was present) was exceedingly delighted with the sight of the model, and passed some time in questioning the divers material things concerning it, and demanded whether I could build the great ship in all parts like the same; for I will, says His Majesty, compare them together when she shall be finished. Then the Lord Admiral commanded me to tell His Majesty the story of the Three Ravens[23] I had seen at Lisbon, in St. Vincent's Church; which I did as well as I could, with my best expressions, though somewhat daunted at first at His Majesty's presence, having never before spoken before any King."

Before, however, he could accomplish his purpose, Pett was overtaken by misfortunes. His enemies, very likely seeing with spite the favour with which he had been received by men in high position, stirred up an agitation against him. There may, and there very probably was, a great deal of jobbery going on in the dockyards. It was difficult, under the system which prevailed, to have any proper check upon the expenditure for the repair and construction of ships. At all events, a commission was appointed for the purpose of inquiring into the abuses and misdemeanors of those in office; and Pett's enemies took care that his past proceedings should be thoroughly overhauled,—together with those of Sir Robert Mansell, then Treasurer to the Navy; Sir John Trevor, surveyor; Sir Henry Palmer, controller; Sir Thomas Bluther, victualler; and many others.

While the commission was still sitting and holding what Pett calls their "malicious proceedings," he was able to lay the keel of his new great ship upon the stocks in the dock at Woolwich on the 20th of October, 1608. He had a clear conscience, for his hands were clean. He went on vigorously with his work, though he knew that the inquisition against him was at its full height. His enemies reported that he was "no artist, and that he was altogether insufficient to perform such a service" as that of building his great ship. Nevertheless, he persevered, believing in the goodness of his cause. Eventually, he was enabled to turn the tables upon his accusers, and to completely justify himself in all his transactions with the king, the Lord Admiral, and the public officers, who were privy to all his transactions. Indeed, the result of the enquiry was not only to cause a great trouble and expense to all the persons accused, but, as Pett says in his Memoir, "the Government itself of that royal office was so shaken and disjoined as brought almost ruin upon the whole Navy, and a far greater charge to his Majesty in his yearly expense than ever was known before."[24]

In the midst of his troubles and anxieties, Pett was unexpectedly cheered with the presence of his "Master" Prince Henry, who specially travelled out of his way from Essex to visit him at Woolwich, to see with his own eyes what progress he was making with the great ship. After viewing the dry dock, which had been constructed by Pett, and was one of the first, if not the very first in England,—his Highness partook of a banquet which the shipbuilder had hastily prepared for him in his temporary lodgings.

One of the circumstances which troubled Pett so much at this time, was the strenuous opposition of the other shipbuilders to his plans of the great ship. There never had been such a frightful innovation. The model was all wrong. The lines were detestable. The man who planned the whole thing was a fool, a "cozener" of the king, and the ship, suppose it to be made, was "unfit for any other use but a dung-boat!" This attack upon his professional character weighed very heavily upon his mind.

He determined to put his case in a staightforward manner before the Lord High Admiral. He set down in writing in the briefest manner everything that he had done, and the plots that had been hatched against him; and beseeched his lordship, for the honour of the State, and the reputation of his office, to cause the entire matter to be thoroughly investigated "by judicious and impartial persons." After a conference with Pett, and an interview with his Majesty, the Lord High Admiral was authorised by the latter to invite the Earls of Worcester and Suffolk to attend with him at Woolwich, and bring all the accusers of Pett's design of the great ship before them for the purpose of examination, and to report to him as to the actual state of affairs. Meanwhile Pett's enemies had been equally busy. They obtained a private warrant from the Earl of Northampton[25] to survey the work; "which being done," says Pett, "upon return of the insufficiency of the same under their hands, and confirmation by oath, it was resolved amongst them I should be turned out, and for ever disgraced."

But the lords appointed by the King now interfered between Pett and his adversaries. They first inspected the ship, and made a diligent survey of the form and manner of the work and the goodness of the materials, and then called all the accusers before them to hear their allegations. They were examined separately. First, Baker the master shipbuilder was called. He objected to the size of the ship, to the length, breadth, depth, draught of water, height of jack, rake before and aft, breadth of the floor, scantling of the timber, and so on. Then another of the objectors was called; and his evidence was so clearly in contradiction to that which had already been given, that either one or both must be wrong. The principal objector, Captain Waymouth, next gave his evidence; but he was able to say nothing to any purpose, except giving their lordships "a long, tedious discourse of proportions, measures, lines, and an infinite rabble of idle and unprofitable speeches, clean from the matter."

The result was that their lordships reported favourably of the design of the ship, and the progress which had already been made.

The Earl of Nottingham interposed his influence; and the King himself, accompanied by the young Prince, went down to Woolwich, and made a personal examination.[26] A great many witnesses were again examined, twenty-four on one side, and twenty-seven on the other. The King then carefully examined the ship himself: "the planks, the tree-nails, the workmanship, and the cross-grained timber." "The cross-grain," he concluded, "was in the men and not in the timber." After all the measurements had been made and found correct, "his Majesty," says Pett, "with a loud voice commanded the measurers to declare publicly the very truth; which when they had delivered clearly on our side, all the whole multitude heaved up their hats, and gave a great and loud shout and acclamation. And then the Prince, his Highness, called with a high voice in these words: 'Where be now these perjured fellows that dare thus abuse his Majesty with these false accusations? Do they not worthily deserve hanging?'"

Thus Pett triumphed over all his enemies, and was allowed to finish the great ship in his own way. By the middle of September 1610, the vessel was ready to be "strucken down upon her ways"; and a dozen of the choice master carpenters of his Majesty's navy came from Chatham to assist in launching her. The ship was decorated, gilded, draped, and garlanded; and on the 24th the King, the Queen, and the Royal family came from the palace at Theobald's to witness the great sight. Unfortunately, the day proved very rough; and it was little better than a neap tide. The ship started very well, but the wind "overblew the tide"; she caught in the dock-gates, and settled hard upon the ground, so that there was no possibility of launching her that day.

This was a great disappointment. The King retired to the palace at Greenwich, though the Prince lingered behind. When he left, he promised to return by midnight, after which it was proposed to make another effort to set the ship afloat. When the time arrived, the Prince again made his appearance, and joined the Lord High Admiral, and the principal naval officials. It was bright moonshine. After midnight the rain began to fall, and the wind to blow from the southwest. But about two o'clock, an hour before high water, the word was given to set all taut, and the ship went away without any straining of screws and tackles, till she came clear afloat into the midst of the Thames. The Prince was aboard, and amidst the blast of trumpets and expressions of joy, he performed the ceremony of drinking from the great standing cup, and throwing the rest of the wine towards the half-deck, and christening the ship by the name of the Prince Royal.[27]

The dimensions of the ship may be briefly described. Her keel was 114 feet long, and her cross-beam 44 feet. She was of 1400 tons burthen, and carried 64 pieces of great ordnance. She was the largest ship that had yet been constructed in England.

The Prince Royal was, at the time she was built, considered one of the most wonderful efforts of human genius. Mr. Charnock, in his 'Treatise on Marine Architecture,' speaks of her as abounding in striking peculiarities. Previous to the construction of this ship, vessels were built in the style of the Venetian galley, which although well adapted for the quiet Mediterranean, were not suited for the stormy northern ocean. The fighting ships also of the time of Henry VIII. and Elizabeth were too full of "top-hamper" for modern navigation. They were oppressed by high forecastles and poops. Pett struck out entirely new ideas in the build and lines of his new ship; and the course which he adopted had its effect upon all future marine structures. The ship was more handy, more wieldy, and more convenient. She was unquestionably the first effort of English ingenuity in the direction of manageableness and simplicity. "The vessel in question," says Charnock, "may be considered the parent of the class of shipping which continues in practice even to the present moment."

It is scarcely necessary to pursue in detail the further history of Phineas Pett. We may briefly mention the principal points. In 1612, the Prince Royal was appointed to convey the Princess Elizabeth and her husband, The Palsgrave, to the Continent. Pett was on board the ship, and found that "it wrought exceedingly well, and was so yare of conduct that a foot of helm would steer her." While at Flushing, "such a multitude of people, men, women, and children, came from all places in Holland to see the ship, that we could scarce have room to go up and down till very night."

About the 27th of March, 1616, Pett bargained with Sir Waiter Raleigh to build a vessel of 500 tons,[28] and received 500L. from him on account. The King, through the interposition of the Lord Admiral, allowed Pett to lay her keel on the galley dock at Woolwich. In the same year he was commissioned by the Lord Zouche, now Lord Warden of the Cinque Ports, to construct a pinnace of 40 tons, in respect of which Pett remarks, "towards the whole of the hull of the pinnace, and all her rigging, I received only 100L. from the Lord Zouche, the rest Sir Henry Mainwaring (half-brother to Raleigh) cunningly received on my behalf, without my knowledge, which I never got from him but by piecemeal, so that by the bargain I was loser 100L. at least."

Pett fared much worse at the hands of Raleigh himself. His great ship, the Destiny, was finished and launched in December, 1616. "I delivered her to him," says Pett, "on float, in good order and fashion; by which business I lost 700L., and could never get any recompense at all for it; Sir Walter going to sea and leaving me unsatisfied."[29] Nor was this the only loss that Pett met with this year. The King, he states, "bestowed upon me for the supply of my present relief the making of a knight-baronet," which authority Pett passed to a recusant, one Francis Ratcliffe, for 700L.; but that worthy defrauded him, so that he lost 30L. by the bargain.

Next year, Pett was despatched by the Government to the New Forest in Hampshire, "where," he says, "one Sir Giles Mompesson[30] had made a vast waste in the spoil of his Majesty's timber, to redress which I was employed thither, to make choice out of the number of trees he had felled of all such timber as was useful for shipping, in which business I spent a great deal of time, and brought myself into a great deal of trouble." About this period, poor Pett's wife and two of his children lay for some time at death's door. Then more enquiries took place into the abuses of the dockyards, in which it was sought to implicate Pett. During the next three years (1618-20) he worked under the immediate orders of the Commissioners in the New Dock at Chatham.

In 1620, Pett's friend Sir Robert Mansell was appointed General of the Fleet destined to chastise the Algerine pirates, who still continued their depredations on the shipping in the Channel, and the King thereupon commissioned Pett to build with all dispatch two pinnaces, of 120 and 80 tons respectively. "I was myself," he says, "to serve as Captain in the voyage"—being glad, no doubt, to escape from his tormentors. The two pinnaces were built at Ratcliffe, and were launched on the 16th and 18th of October, 1620. On the 30th, Pett sailed with the fleet, and after driving the pirates out of the Channel, he returned to port after an absence of eleven months.

His enemies had taken advantage of his absence from England to get an order for the survey of the Prince Royal, his masterpiece; the result of which was, he says, that "they maliciously certified the ship to be unserviceable, and not fit to continue—that what charges should be bestowed upon her would be lost." Nevertheless, the Prince Royal was docked, and fitted for a voyage to Spain. She was sent thither with Charles Prince of Wales and the Duke of Buckingham, the former going in search of a Spanish wife. Pett, the builder of the ship, was commanded to accompany the young Prince and the Duke.

The expedition sailed on the 24th of August, 1623, and returned on the 14th of October. Pett was entertained on board the Prince Royal, and rendered occasional services to the officers in command, though nothing of importance occurred during the voyage.

The Prince of Wales presented him with a valuable gold chain as a reward for his attendance. In 1625, Pett, after rendering many important services to the Admiralty, was ordered again to prepare the Prince Royal for sea. She was to bring over the Prince of Wales's bride from France. While the preparations were making for the voyage, news reached Chatham of the death of King

James. Pett was afterwards commanded to go forward with the work of preparing the Prince Royal, as well as the whole fleet, which was intended to escort the French Princess, or rather the Queen, to England. The expedition took place in May, and the young Queen landed at Dover on the 12th of that month.

Pett continued to be employed in building and repairing ships, as well as in preparing new designs, which he submitted to the King and the Commissioners of the Navy. In 1626, he was appointed a joint commissioner, with the Lord High Admiral, the Lord Treasurer Marlborough, and others, "to enquire into certain alleged abuses of the Navy, and to view the state thereof, and also the stores thereof," clearly showing that he was regaining his old position. He was also engaged in determining the best mode of measuring the tonnage of ships.[31] Four years later he was again appointed a commissioner for making "a general survey of the whole navy at Chatham." For this and his other services the King promoted Pett to be a principal officer of the Navy, with a fee of 200L. per annum. His patent was sealed on the 16th of January, 1631. In the same year the King visited Woolwich to witness the launching of the Vanguard, which Pett had built; and his Majesty honoured the shipwright by participating in a banquet at his lodgings.

From this period to the year 1637, Pett records nothing of particular importance in his autobiography. He was chiefly occupied in aiding his son Peter—who was rapidly increasing his fame as a shipwright—in repairing and building first-class ships of war. As Pett had, on an early occasion in his life, prepared a miniature ship for Prince Henry, eldest son of James I., he now proceeded to prepare a similar model for the Prince of Wales, the King's eldest son, afterwards Charles II. This model was presented to the Prince at St. James's, "who entertained it with great joy, being purposely made to disport himself withal." On the next visit of his Majesty to Woolwich, he inspected the progress made with the Leopard, a sloop-of-war built by Peter Pett. While in the hold of the vessel, the King called Phineas to one side, and told him of his resolution to have a great new ship built, and that Phineas must be the builder. This great new ship was The Sovereign of the Seas, afterwards built by Phineas and Peter Pett. Some say that the model was prepared by the latter; but Phineas says that it was prepared by himself, and finished by the 29th of October, 1634. As a compensation for his services, his Majesty renewed his pension of 40L. (which had been previously stopped), with orders for all the arrears due upon it to be paid.

To provide the necessary timber for the new ship, Phineas and his son went down into the North to survey the forests. They went first by water to Whitby; from thence they proceeded on horseback to Gisborough and baited; then to Stockton, where they found but poor entertainment, though they lodged with the Mayor, whose house "was only a mean thatched cottage!" Middlesborough and the great iron district of the North had not yet come into existence.

Newcastle, already of some importance, was the principal scene of their labours. The timber for the new ship was found in Chapley Wood and Bracepeth Park. The gentry did all they could to facilitate the object of Pett. On his journey homewards (July, 1635), he took Cambridge on his way, where, says he, "I lodged at the Falcon, and visited Emmanuel College, where I had been a scholar in my youth."

The Sovereign of the Seas was launched on the 12th of October, 1637, having been about two years in building. Evelyn in his diary says of the ship (19th July, 1641):—"We rode to Rochester and Chatham to see the Soveraigne, a monstrous vessel so called, being for burthen, defence, and ornament, the richest that ever spread cloth before the wind. She carried 100 brass cannon, and was 1600 tons, a rare sailer, the work of the famous Phineas Pett." Rear-Admiral Sir William Symonds says that she was afterwards cut down, and was a safe and fast ship.[32]

The Sovereign continued for nearly sixty years to be the finest ship in the English service. Though frequently engaged in the most injurious occupations, she continued fit for any services which the exigencies of the State might require. She fought all through the wars of the Commonwealth; she was the leading ship of Admiral Blake, and was in all the great naval engagements with France and Holland. The Dutch gave her the name of The Golden Devil. In the last fight between the English and French, she encountered the Wonder of the World, and so warmly plied the French Admiral, that she forced him out of his three-decked wooden castle, and chasing the Royal Sun, before her, forced her to fly for shelter among the rocks, where she became a prey to lesser vessels, and was reduced to ashes. At last, in the reign of William III., the Sovereign became leaky and defective with age; she was laid up at Chatham, and being set on fire by negligence or accident, she burnt to the water's edge.

To return to the history of Phineas Pett. As years approached, he retired from office, and "his loving son," as he always affectionately designates Peter, succeeded him as principal shipwright, Charles I. conferring upon him the honour of knighthood. Phineas lived for ten years after the Sovereign of the Seas was launched. In the burial register of the parish of Chatham it is recorded, "Phineas Pett, Esqe. and Capt., was buried 21st August, 1647."[33]

Sir Peter Pett was almost as distinguished as his father. He was the builder of the first frigate, The Constant Warwick. Sir William Symonds says of this vessel:—"She was an incomparable sailer, remarkable for her sharpness and the fineness of her lines; and many were built like her." Pett "introduced convex lines on the immersed part of the hull, with the studding and sprit sails; and, in short, he appears to have fully deserved his character of being the best ship architect of his time."[34] Sir Peter Pett's monument in Deptford Old Church fully records his services to England's naval power.

The Petts are said to have been connected with shipbuilding in the Thames for not less than 200 years. Fuller, in his 'Worthies of England,' says of them—"I am credibly informed that that mystery of shipwrights for some descents hath been preserved faithfully in families, of whom the Petts about Chatham are of singular regard. Good success have they with their skill, and carefully keep so precious a pearl, lest otherwise amongst many friends some foes attain unto it."

The late Peter Bolt, member for Greenwich, took pride in being descended from the Petts; but so far as we know, the name itself has died out. In 1801, when Charnock's 'History of Marine Architecture' was published, Mr. Pett, of Tovil, near Maidstone, was the sole representative of the family.

Footnotes for Chapter I.

[1] This was not the first voyage of a steamer between England and America. The Savannah made the passage from New York to Liverpool as early as 1819; but steam was only used occasionally during the voyage, In 1825, the Enterprise, with engines by Maudslay, made the voyage from Falmouth to Calcutta in 113 days; and in 1828, the Curacoa made the voyage between Holland and the Dutch West Indies. But in all these cases, steam was used as an auxiliary, and not as the one essential means of propulsion, as in the case of the Sirius and the Great Western, which were steam voyages only.

[2] "In 1862 the steam tonnage of the country was 537,000 tons; in 1872, it was 1,537,000 tons; and in 1882, it had reached 3,835,000 tons."—Mr. Chamberlain's speech, House of Commons, 19th May, 1884.

[3] The last visit of the plague was in 1665.

[4] Roll of Edward the Third's Fleet. Cotton's Library, British Museum.

[5] Charnock's History Of Marine Architecture, ii. 89.

[6] State Papers. Henry VIII. Nos. 3496, 3616, 4633. The principal kinds of ordnance at that time were these:—The "Apostles," so called from the head of an Apostle which they bore; "Curtows," or "Courtaulx"; "Culverins" and "Serpents"; "Minions," and "Potguns"; "Nurembergers," and "Bombards" or mortars.

[7] The sum of all costs of the Harry Grace de Dieu and three small galleys, was 7708L. 5s. 3d. (S.P.O. No. 5228, Henry VIII.)

[8] Charnock, ii. 47 (note).

[9] Macpherson, Annals of Commerce, ii. 126.

[10] The Huguenots: their Settlements, Churches, and Industries, in England and Ireland, ch. iv.

[11] Macpherson, Annals of Commerce, ii. 156.

[12] Ibid. ii. 85.

[13] Picton's Selections from the Municipal Archives and Records of Liverpool, p. 90. About a hundred years later, in 1757, the gross customs receipts of Liverpool had increased to 198,946L.; whilst those of Bristol were as much as

351,211L. In 1883, the amount of tonnage of Liverpool, inwards and outwards, was 8,527,531 tons, and the total dock revenue for the year was 1,273,752L.!

[14] There were not only Algerine but English pirates scouring the seas. Keutzner, the German, who wrote in Elizabeth's reign, said, "The English are good sailors and famous pirates (sunt boni nautae et insignis pyratae)." Roberts, in his Social History of the Southern Counties (p. 93), observes, "Elizabeth had employed many English as privateers against the Spaniard. After the war, many were loth to lead an inactive life. They had their commissions revoked, and were proclaimed pirates. The public looked upon them as gallant fellows; the merchants gave them underhand support; and even the authorities in maritime towns connived at the sale of their plunder. In spite of proclamations, during the first five years after the accession of James I., there were continual complaints. This lawless way of life even became popular. Many Englishmen furnished themselves with good ships and scoured the seas, but little careful whom they might plunder." It was found very difficult to put down piracy. According to Oliver's History of the city of Exeter, not less than "fifteen sail of Turks" held the English Channel, snapping up merchantmen, in the middle of the seventeenth century! The harbours in the south-west were infested by Moslem pirates, who attacked and plundered the ships, and carried their crews into captivity. The loss, even to an inland port like Exeter, in ships, money, and men, was enormous.

[15] Naval Tracts, p. 294.

[16] This poem is now very rare. It is not in the British Museum.

[17] There are three copies extant of the autobiography, all of which are in the British Museum. In the main, they differ but slightly from each other. Not one of them has been published in extenso. In December, 1795, and in February, 1796, Dr. Samuel Denne communicated to the Society of Antiquaries particulars of two of these MSS., and subsequently published copious extracts from them in their transactions (Archae. xii. anno 1796), in a very irregular and careless manner. It is probable that Dr. Denne never saw the original manuscript, but only a garbled copy of it. The above narrative has been taken from the original, and collated with the documents in the State Paper Office.

[18] See, for instance, the Index to the Journals of Records of the Corporation of the City of London (No. 2, p. 346, 15901694) under the head of "Sir Walter Raleigh." There is a document dated the 15th November, 1593, in the 35th of Elizabeth, which runs as follows:—"Committee appointed on behalf of such of the City Companies as have ventured in the late Fleet set forward by Sir Walter Raleigh, Knight, and others, to join with such honourable personages as the Queen hath appointed, to take a perfect view of all such goods, prizes, spices, jewels, pearls, treasures, &c., lately taken in the Carrack, and to make sale and division (Jor. 23, p. 156). Suit to be made to the Queen and Privy Council for the buying of the goods, &c., lately taken at sea in the Carrack; a committee appointed to take order accordingly; the benefit or loss arising thereon to be divided and borne between the Chamber [of the Corporation of the City] and the Companies that adventured (157). The several Companies that adventured at sea with Sir Waiter Raleigh to accept so much of the goods taken in the Carrack to the value of 12,000L. according to the Queen's offer. A committee appointed to acquaint the Lords of the Council with the City's acceptance thereof (167). Committee for sale of the Carrack goods appointed (174). Bonds for sale to be sealed (196).... Committee to audit accounts of a former adventure (224 b.)."

[19] There were three sisters in all, the eldest of whom (Abigail) fell a victim to the cruelty of Nunn, who struck her across the head with the fire-tongs, from the effects of which she died in three days. Nunn was tried and convicted of manslaughter. He died shortly after. Mrs. Nunn, Phineas's mother, was already dead.

[20] It would seem, from a paper hereafter to be more particularly referred to, that the government encouraged the owners of ships and others to clear the seas of these pirates, agreeing to pay them for their labours. In 1622, Pett fitted out an expedition against these pests of navigation, but experienced some difficulty in getting his expenses repaid.

[21] See grant S.P.O., 29th May, 1605.

[22] An engraving of this remarkable ship is given in Charnock's History of Marine Architecture, ii. p. 199.

[23] The story of the Three, or rather Two Ravens, is as follows:—The body of St. Vincent was originally deposited at the Cape, which still bears his name, on the Portuguese coast; and his tomb, says the legend, was zealously guarded by a couple of ravens. When it was determined, in the 12th century, to transport the relics of the Saint to the Cathedral of Lisbon, the two ravens accompanied the ship which contained them, one at its stem and the other at its stern. The relics were deposited in the Chapel of St. Vincent, within the Cathedral, and there the two ravens have ever since remained. The monks continued to support two such birds in the cloisters, and till very lately the officials gravely informed the visitor to the Cathedral that they were the identical ravens which accompanied the Saint's relics to their city. The birds figure in the arms of Lisbon.

[24] The evidence taken by the Commissioners is embodied in a voluminous report. State Paper Office, Dom. James I., vol. xli. 1608.

[25] The Earl of Northampton, Privy Seal, was Lord Warden of the Cinque Ports; hence his moving in the matter. Pett says he was his "most implacable enemy." It is probable that the earl was jealous of Pett, because he had received his commission to build the great ship directly from the sovereign, without the intervention of his lordship.

[26] This Royal investigation took place at Woolwich on the 8th May, 1609. The State Paper Office contains a report of the same date, most probably the one presented to the King, signed by six ship-builders and Captain Waymouth, and counter signed by Northampton and four others. The Report is headed "The Prince Royal: imperfections found upon view of the new work begun at Woolwich." It would occupy too much space to give the results here.

[27] Alas! for the uncertainties of life! This noble young prince—the hope of England and the joy of his parents, from whom such great things were anticipated—for he was graceful, frank, brave, active, and a lover of the sea,—was seized with a serious illness, and died in his eighteenth year, on the 16th November, 1612.

[28] Pett says she was to be 500 tons, but when he turned her out her burthen was rated at 700 tons.

[29] This conduct of Raleigh's was the more inexcusable, as there is in the State Paper Office a warrant dated 16th Nov., 1617, for the payment to Pett of 700 crowns "for building the new ship, the Destiny of London, of 700 tons burthen." The least he could have done was to have handed over to the builder his royal and usual reward. In the

above warrant, by the way, the title "our well-beloved subject," the ordinary prefix to such grants, has either been left blank or erased (it is difficult to say which), but was very significant of the slippery footing of Raleigh at Court.

[30] Sir Giles Overreach, in the play of "A new way to pay old debts," by Philip Massinger. It was difficult for the poet, or any other person, to libel such a personage as Mompesson.

[31] Pett's method is described in a paper contained in the S.P.O., dated 21st Oct., 1626. The Trinity Corporation adopted his method.

[32] Memoirs of the Life and Services of Rear-Admiral Sir William Symonds, Kt., p. 94.

[33] Pett's dwelling-house at Rochester is thus described in an anonymous history of that town (p. 337, ed. 1817):—"Beyond the Victualling Office, on the same side of the High Street, at Rochester, is an old mansion, now occupied by a Mr. Morson, an attorney, which formerly belonged to the Petts, the celebrated ship-builders. The chimney-piece in the principal room is of wood, curiously carved, the upper part being divided into compartments by caryatydes. The central compartment contains the family arms, viz., Or, on a fesse, gu., between three pellets, a lion passant gardant of the field. On the back of the grate is a cast of Neptune, standing erect in his car, with Triton blowing conches, &c., and the date 1650."

[34] Symonds, Memoirs of Life and Services, 94.

CHAPTER II.

FRANCIS PETTIT SMITH: PRACTICAL INTRODUCER OF THE SCREW PROPELLER.

"The spirit of Paley's maxim that 'he alone discovers who proves,' is applicable to the history of inventions and discoveries; for certainly he alone invents to any good purpose, who satisfies the world that the means he may have devised have been found competent to the end proposed."—Dr. Samuel Brown.

"Too often the real worker and discoverer remains unknown, and an invention, beautiful but useless in one age or country, can be applied only in a remote generation, or in a distant land.

Mankind hangs together from generation to generation; easy labour is but inherited skill; great discoveries and inventions are worked up to by the efforts of myriads ere the goal is reached."—H. M. Hyndman.

Though a long period elapsed between the times of Phineas Pett and "Screw" Smith, comparatively little improvement had been effected in the art of shipbuilding. The Sovereign of the Seas had not been excelled by any ship of war built down to the end of last century.[1] At a comparatively recent date, ships continued to be built of timber and plank, and impelled by sails and oars, as they had been for thousands of years before.

But this century has witnessed many marvellous changes. A new material of construction has been introduced into shipbuilding, with entirely new methods of propulsion. Old things have been displaced by new; and the magnitude of the results has been extraordinary. The most important changes have been in the use of iron and steel instead of wood, and in the employment of the steam-engine in impelling ships by the paddle or the screw.

So long as timber was used for the construction of ships, the number of vessels built annually, especially in so small an island as Britain, must necessarily have continued very limited. Indeed, so little had the cultivation of oak in Great Britain been attended to, that all the royal forests could not have supplied sufficient timber to build one line-of-battle ship annually; while for the mercantile marine, the world had to be ransacked for wood, often of a very inferior quality.

Take, for instance, the seventy-eight gun ship, the Hindostan, launched a few years ago. It would have required 4200 loads of timber to build a ship of that description, and the growth of the timber would have occupied seventy acres of ground during eighty years.[2] It would have needed something like 800,000 acres of land on which to grow the timber for the ships annually built in this country for commercial purposes. And timber ships are by no means lasting. The average durability of ships of war employed in active service, has been calculated to be about thirteen years, even when built of British oak.

Indeed, years ago, the building of shipping in this country was much hindered by the want of materials.

The trade was being rapidly transferred to Canada and the United States. Some years since, an American captain said to an Englishman, Captain Hall, when in China, "You will soon have to come to our country for your ships: your little island cannot grow wood enough for a large marine." "Oh!" said the Englishman, "we can build ships of iron!" "Iron?" replied the American in surprise, "why, iron sinks; only wood can float!" "Well! you will find I am right." The prophecy was correct. The Englishman in question has now a fleet of splendid iron steamers at sea.

The use of iron in shipbuilding had small beginnings, like everything else. The established prejudice—that iron must necessarily sink in water—long continued to prevail against its employment. The first iron vessel was built and launched about a hundred years since by John Wilkinson, of Bradley Forge, in Staffordshire. In a letter of his, dated the 14th July, 1787, the original of which we have seen, he writes: "Yesterday week my iron boat was launched. It answers all my expectations, and has convinced the unbelievers, who were 999 in 1000. It will be only a nine days' wonder, and afterwards a Columbus's egg." It was, however, more than a nine days' wonder; for wood long continued to be thought the only material capable of floating.

Although Wilkinson's iron vessels continued to ply upon the Severn, more than twenty years elapsed before another shipbuilder ventured to follow his example. But in 1810, Onions and Son, of Brosely, built several iron vessels, also for use upon the Severn. Then, in 1815, Mr. Jervons, of Liverpool, built a small iron boat for use on the Mersey. Six years later, in 1821, Mr. Aaron Manby designed an iron steam vessel, which was built at the Horsley Company's Works, in Staffordshire. She sailed from London to Havre a few years later, under the command of Captain (afterwards Sir Charles) Napier, RN. She was freighted with a cargo of linseed and iron castings, and went up the Seine to Paris. It was some time, however, before iron came into general use. Ten years later, in 1832, Maudslay and Field built four iron vessels for the East India Company. In the course of about twenty years, the use of iron became general, not only for ships of war, but for merchant ships plying to all parts of the world.

When ships began to be built of iron, it was found that they could be increased without limit, so long as coal, iron, machinery, and strong men full of skill and industry, were procurable. The trade in shipbuilding returned to Britain, where iron ships are now made and exported in large numbers; the mercantile marine of this country exceeding in amount and tonnage that of all the other countries of the world put together. The "wooden walls"[3] of England exist no more, for iron has superseded wood. Instead of constructing vessels from the forest, we are now digging new navies out of the bowels of the earth, and our "walls," instead of wood, are now of iron and steel.

The attempt to propel ships by other means than sails and oars went on from century to century, and did not succeed until almost within our own time. It is said that the Roman army under Claudius Codex was transported into Sicily in boats propelled by wheels moved by oxen. Galleys, propelled by wheels in paddles, were afterwards attempted. The Harleian MS. contains an Italian book of sketches, attributed to the 15th century, in which there appears a drawing of a paddle-boat, evidently intended to be worked by men. Paddle-boats, worked by horse-power, were also tried. Blasco Garay made a supreme effort at Barcelona in 1543. His vessel was propelled by a paddle-wheel on each side, worked by forty men. But nothing came of the experiment.

Many other efforts of a similar kind were made,—by Savery among others,[4]—until we come down to Patrick Miller, of Dalswinton, who, in 1787, invented a double-hulled boat, which he caused to be propelled on the Firth of Forth by men working a capstan which drove the paddles on each side. The men soon became exhausted, and on Miller mentioning the subject to

William Symington, who was then exhibiting his road locomotive in Edinburgh, Symington at once said, "Why don't you employ steam-power?"

There were many speculations in early times as to the application of steam-power for propelling vessels through the water. David Ramsay in 1618, Dr. Grant in 1632, the Marquis of Worcester in 1661, were among the first in England to publish their views upon the subject. But it is probable that Denis Papin, the banished Hugnenot physician, for some time Curator of the Royal Society, was the first who made a model steam-boat. Daring his residence in England, he was elected Professor of Mathematics in the University of Marburg. It was while at that city that he constructed, in 1707, a small steam-engine, which he fitted in a boat—une petite machine d'un, vaisseau a roues—and despatched it to England for the purpose of being tried upon the Thames. The little vessel never reached England. At Munden, the boatmen on the River Weser, thinking that, if successful, it would destroy their occupation, seized the boat, with its machine, and barbarously destroyed it. Papin did not repeat his experiment, and died a few years later.

The next inventor was Jonathan Hulls, of Campden, in Gloucestershire. He patented a steamboat in 1736, and worked the paddle-wheel placed at the stern of the vessel by means of a Newcomen engine. He tried his boat on the River Avon, at Evesham, but it did not succeed, and the engine was taken on shore again. A local poet commemorated his failure in the following lines, which were remembered long after his steamboat experiment had been forgotten:—

"Jonathan Hull,
With his paper skull,
Tried hard to make a machine
That should go against wind and tide;
But he, like an ass,
Couldn't bring it to pass,
So at last was ashamed to be seen."

Nothing of importance was done in the direction of a steam-engine able to drive paddles, until the invention by James Watt, in 1769, of his double-acting engine—the first step by which steam was rendered capable of being successfully used to impel a vessel. But Watt was indifferent to taking up the subject of steam navigation, as well as of steam locomotion. He refused many invitations to make steam-engines for the propulsion of ships, preferring to confine himself to his "regular established trade and manufacture," that of making condensing steam-engines, which had become of great importance towards the close of his life.

Two records exist of paddle-wheel steamboats having been early tried in France—one by the Comte d'Auxiron and M. Perrier in 1774, the other by the Comte de Jouffroy in 1783—but the notices of their experiments are very vague, and rest on somewhat doubtful authority.

The idea, however, had been born, and was not allowed to die. When Mr. Miller of Dalswinton had revived the notion of propelling vessels by means of paddle-wheels, worked, as Savery had before worked them, by means of a capstan placed in the centre of the vessel, and when he complained to Symington of the fatigue caused to the men by working the capstan, and Symington had suggested the use of steam, Mr. Miller was impressed by the idea, and proceeded to order a steam-engine for the purpose of trying the experiment. The boat was built at Edinburgh, and removed to Dalswinton Lake. It was there fitted with Symington's steam-engine, and first tried with success on the 14th of October, 1788, as has been related at length in Mr. Nasmyth's 'Autobiography.' The experiment was repeated with even greater success in the charlotte Dundas in 1801, which was used to tow vessels along the Forth and Clyde Canal, and to bring ships up the Firth of Forth to the canal entrance at Grangemouth.

The progress of steam navigation was nevertheless very slow. Symington's experiments were not renewed. The Charlotte Dundas was withdrawn from use, because of the supposed injury to the banks of the Canal, caused by the swell from the wheel. The steamboat was laid up in a creek at Bainsford, where it went to ruin, and the inventor himself died in poverty. Among those who inspected the vessel while at work were Fulton, the American artist, and Henry Bell, the Glasgow engineer. The former had already occupied himself with model steamboats, both at Paris and in London; and in 1805 he obtained from Boulton and Watt, of Birmingham, the steam-engine required for propelling his paddle steamboat on the Hudson. The Clermont was first started in August, 1807, and attained a speed of nearly five miles an hour. Five years later, Henry Bell constructed and tried his first steamer on the Clyde.

It was not until 1815 that the first steamboat was seen on the Thames. This was the Richmond packet, which plied between London and Richmond. The vessel was fitted with the first marine engine Henry Maudslay ever made. During the same year, the Margery, formerly employed on the Firth of Forth, began plying between Gravesend and London; and the Thames, formerly the Argyll, came round from the Clyde, encountering rough seas, and making the voyage of 758 miles in five days and two hours. This was thought extraordinarily rapid—though the voyage of about 3000 miles, from Liverpool to New York, can now be made in only about two days' more time.

In nearly all seagoing vessels, the Paddle has now almost entirely given place to the Screw. It was long before this invention was perfected and brought into general use. It was not the production of one man, but of several generations of mechanical inventors. A perfected invention does not burst forth from the brain like a poetic thought or a fine resolve. It has to be initiated, laboured over, and pursued in the face of disappointments, difficulties, and discouragements.

Sometimes the idea is born in one generation, followed out in the next, and perhaps perfected in the third. In an age of progress, one invention merely paves the way for another. What was the wonder of yesterday, becomes the common and unnoticed thing of to-day.

The first idea of the screw was thrown out by James Watt more than a century ago. Matthew Boulton, of Birmingham, had proposed to move canal boats by means of the steam-engine; and Dr. Small, his friend, was in communication with James Watt, then residing at Glasgow, on the subject. In a letter from Watt to Small, dated the 30th September, 1770, the former, after speaking of the condenser, and saying that it cannot be dispensed with, proceeds: "Have you ever considered a spiral oar for that purpose [propulsion of canal boats], or are you for two wheels?" Watt added a pen-and-ink drawing of his spiral oar, greatly resembling the form of screw afterwards patented. Nothing, however, was actually done, and the idea slept.

It was revived again in 1785, by Joseph Bramah, a wonderful projector and inventor.[5] He took out a patent, which included a rotatory steam-engine, and a mode of propelling vessels by means either of a paddle-wheel or a "screw propeller." This propeller was "similar to the fly of a smoke-jack"; but there is no account of Bramah having practically tried this method of propulsion.

Austria, also, claims the honour of the invention of the screw steamer. At Trieste and Vienna are statues erected to Joseph Ressel, on whose behalf his countrymen lay claim to the invention; and patents for some sort of a screw date back as far as 1794.

Patents were also taken out in England and America—by W. Lyttleton in 1794; by E. Shorter in 1799; by J. C. Stevens, of New Jersey, in 1804; by Henry James in 1811—but nothing practical was accomplished. Richard Trevethick, the anticipator of many things, also took out a patent in 1815, and in it he describes the screw propeller with considerable minuteness. Millington, Whytock, Perkins, Marestier, and Brown followed, with no better results.

The late Dr. Birkbeck, in a letter addressed to the 'Mechanics' Register,' in the year 1824, claimed that John Swan, of 82, Mansfield Street, Kingsland Road, London, was the practical inventor of the screw propeller. John Swan was a native of Coldingham, Berwickshire. He had removed to London, and entered the employment of Messrs. Gordon, of Deptford. Swan fitted up a boat with his propeller, and tried it on a sheet of water in the grounds of Charles Gordon, Esq., of Dulwich Hill. "The velocity and steadiness of the motion," said Dr. Birkbeck in his letter, "so far exceeded that of the same model when impelled by paddle-wheels driven by the same spring, that I could not doubt its superiority; and the stillness of the water was such as to give the vessel the appearance of being moved by some magical power."

Then comes another claimant—Mr. Robert Wilson, then of Dunbar (not far from Coldingham), but afterwards of the Bridgewater Foundry, Patricroft. In his pamphlet, published a few years ago, he states that he had long considered the subject, and in 1827 he made a small model, fitted with "revolving skulls," which he tried on a sheet of water in the presence of the Hon. Capt. Anthony Maitland, son of the Earl of Lauderdale. The experiment was successful—so successful, that when the "stern paddles" were in 1828 used at Leith in a boat twenty-five feet long, with two men to work the machinery, the boat was propelled at an average speed of about ten miles an hour; and the Society of Arts afterwards, in October, 1882, awarded Mr. Wilson their silver medal for the "description, drawing, and models of stern paddles for propelling steamboats, invented by him." The subject was, in 1833, brought by Sir John Sinclair under the consideration of the Board of Admiralty; but the report of the officials (Oliver Lang, Abethell, Lloyd, and Kingston) was to the effect that "the plan proposed (independent of practical difficulties) is objectionable, as it involves a greater loss of power than the common mode of applying the wheels to the side." And here ended the experiment, so far as Mr. Wilson's "stern paddles" were concerned.

It will be observed, from what has been said, that the idea of a screw propeller is a very old one. Watt, Bramah, Trevethick, and many more, had given descriptions of the screw. Trevethick schemed a number of its forms and applications, which have been the subject of many subsequent patents. It has been so with many inventions. It is not the man who gives the first idea of a machine who is entitled to the merit of its introduction, or the man who repeats the idea, and re-repeats it, but the man who is so deeply impressed with the importance of the discovery, that he insists upon its adoption, will take no denial, and at the risk of fame and fortune, pushes through all opposition, and is determined that what he thinks he has discovered shall not perish for want of a fair trial. And that this was the case with the practical introducer of the screw propeller will be obvious from the following statement.

Francis Pettit Smith was born at Hythe, in the county of Kent, in 1808. His father was postmaster of the town, and a person of much zeal and integrity. The boy was sent to school at Ashford, and there received a fair amount of education, under the Rev. Alexander Power. Young Smith displayed no special characteristic except a passion for constructing models of boats. When he reached manhood, he adopted the business of a grazing farmer on Romney Marsh. He afterwards removed to Hendon, north of London, where he had plenty of water on which to try his model boats. The reservoir of the Old Welsh Harp was close at hand—a place famous for its water-birds and wild fowl.

Smith made many models of boats, his experiments extending over many years. In 1834, he constructed a boat propelled by a wooden screw driven by a spring, the performance of which was thought extraordinary. Where he had got his original idea is not known. It was floating about in many minds, and was no special secret. Smith, however, arrived at the conclusion that his method of propelling steam vessels by means of a screw was much superior to paddles—at that time exclusively employed. In the following year, 1835, he constructed a superior model, with which he performed a number of experiments at Hendon. In May 1836, he took out a patent for propelling vessels by means of a screw revolving beneath the water at the stern. He then openly exhibited his invention at the Adelaide Gallery in London. Sir John Barrow, Secretary to the Admiralty, inspected the model, and was much impressed by its action. During the time it was publicly exhibited, an offer was made to purchase the invention for the Pacha of Egypt; but the offer was declined.

At this stage of his operations, Smith was joined by Mr. Wright, banker, and Mr. C. A. Caldwell, who had the penetration to perceive that the invention was one of much promise, and were desirous of helping its introduction to general use. They furnished Smith with the means of constructing a more complete model. In the autumn of 1836, a small steam vessel of 10 tons burthen and six horse-power was built, further to test the advantages of the invention. This boat was fitted with a wooden screw of two whole turns. On the 1st of November the vessel was exhibited to the public on the Paddington Canal, as well as on the Thames, where she continued to ply until the month of September 1837.

During the trips upon the Thames, a happy accident occurred, which first suggested the advantage of reducing the length of the screw. The propeller having struck upon some obstacle in the water, about one-half of the length of the screw was broken off, and it was found that; the vessel immediately shot ahead and attained a much greater speed than before. In consequence of this discovery, a new screw of a single turn was fitted to her, after which she was found to work much better.

Having satisfied himself as to the eligibility of the propeller in smooth water, Mr. Smith then resolved to take his little vessel to the open sea, and breast the winds and the waves. Accordingly, one Saturday in the month of September 1837, he proceeded in his miniature boat, down the river, from Blackwall to Gravesend. There he took a pilot on board, and went on to Ramsgate. He passed through the Downs, and reached Dover in safety. A trial of the vessel's performance was made there in the presence of Mr. Wright, the banker, and Mr. Peake, the civil engineer. From Dover the vessel went on to Folkestone and Hythe, encountering severe weather. Nevertheless, the boat behaved admirably, and attained a speed of over seven miles an hour.

Though the weather had become stormy and boisterous, the little vessel nevertheless set out on her return voyage to London. Crowds of people assembled to witness her departure, and many nautical men watched her progress with solicitude as she steamed through the waves under the steep cliffs of the South Foreland. The courage of the undertaking, and the unexpected good performance of the little vessel, rendered her an object of great interest and excitement as she "screwed" her way along the coast.

The tiny vessel reached her destination in safety. Surely the difficulty of a testing trial, although with a model screw, had at length been overcome. But no! The paddle still possessed the ascendency; and a thousand interests—invested capital, use and wont, and conservative instincts—all stood in the way.

Some years before—indeed, about the time that Smith took out his patent—Captain Ericsson, the Swede, invented a screw propeller. Smith took out his patent in May, 1836; and Ericsson in the following July. Ericsson was a born inventor. While a boy in Sweden, he made saw mills and pumping engines, with tools invented by himself. He learnt to draw, and his mechanical career began. When only twelve years old, he was appointed a cadet in the Swedish corps of mechanical engineers, and in the following year he was put in charge of a section of the Gotha Ship Canal, then under construction. Arrived at manhood, Ericsson went over to England, the great centre of mechanical industry. He was then twenty-three years old. He entered into partnership with John Braithwaite, and with him constructed the Novelty, which took part in the locomotive competition at Rainhill on the 6th October, 1829. The prize was awarded to Stephenson's Rocket on the 14th; but it was acknowledged by The Times of the day that the Novelty was Stephenson's sharpest competitor.

Ericsson had a wonderfully inventive brain, a determined purpose, and a great capacity for work. When a want was felt, he was immediately ready with an invention. The records of the Patent Office show his incessant activity. He invented pumping engines, steam engines, fire engines, and caloric engines. His first patent for a "reciprocating propeller" was taken out in October 1834. To exhibit its action, he had a small boat constructed of only about two feet long. It was propelled by means of a screw; and was shown at work in a circular bath in London. It performed its voyage round the basin at the rate of about three miles an hour. His patent for a "spiral propeller," was taken out in July 1836. This was the invention, to exhibit which he had a vessel constructed, of about 40 feet long, with two propellers, each of 5 feet 3 inches diameter.

This boat, the Francis B. Ogden, proved extremely successful. She moved at a speed of about ten miles an hour. She was able to tow vessels of 140 tons burthen at the rate of seven miles an hour. Perceiving the peculiar and admirable fitness of the screw-propeller for ships of war, Ericsson invited the Lords of the Admiralty to take an excursion in tow of his experimental boat. "My Lords" consented; and the Admiralty barge contained on this occasion, Sir Charles Adam, senior Lord, Sir William Symonds, surveyor, Sir Edward Parry, of Polar fame, Captain Beaufort, hydrographer, and other men of celebrity. This distinguished company embarked at Somerset House, and the little steamer, with her precious charge, proceeded down the river to Limehouse at the rate of about ten miles an hour. After visiting the steam-engine manufactory of Messrs. Seawood, where their Lordships' favourite apparatus, the Morgan paddle-wheel, was in course of construction, they re-embarked, and returned in safety to Somerset House.

The experiment was perfectly successful, and yet the result was disappointment. A few days later, a letter from Captain Beaufort informed Mr. Ericsson that their Lordships had certainly been "very much disappointed with the result of the experiment." The reason for the disappointment was altogether inexplicable to the inventor. It afterwards appeared, however, that Sir William Symonds, then Surveyor to the Navy, had expressed the opinion that "even if the propeller had the power of propelling a vessel, it would be found altogether useless in practice, because the power being applied at the stern, it would be absolutely impossible to make the vessel steer!" It will be remembered that Francis Pettit Smith's screw vessel went to sea in the course of the same year; and not only faced the waves, but was made to steer in a perfectly successful manner.

Although the Lords of the Admiralty would not further encourage the screw propeller of Ericsson, an officer of the United States Navy, Capt. R. F. Stockton, was so satisfied of its success, that after making a single trip in the experimental steamboat from London Bridge to Greenwich, he ordered the inventor to build for him forthwith two iron boats for the United States, with steam machinery and a propeller on the same plan. One of these vessels—the Robert

F. Stockton—seventy feet in length, was constructed by Laird and Co., of Birkenhead, in 1838, and left England for America in April 1839. Capt. Stockton so fully persuaded Ericsson of his probable success in America, that the inventor at once abandoned his professional engagements in England, and set out for the United States. It is unnecessary to mention the further important works of this great engineer.

We may, however, briefly mention that in 1844, Ericsson constructed for the United States Government the Princeton screw steamer—though he was never paid for his time, labour, and expenditure.[6] Undeterred by their ingratitude, Ericsson nevertheless constructed for the same government, when in the throes of civil war, the famous Monitor, the iron-clad cupola vessel, and was similarly rewarded! He afterwards invented the torpedo ship—the Destroyer—the use of which has fortunately not yet been required in sea warfare. Ericsson still lives—constantly planning and scheming—in his house in Beach Street, New York. He is now over eighty years old having been born in 1803. He is strong and healthy. How has he preserved his vigorous constitution? The editor of Scribner gives the answer: "The hall windows of his house are open, winter and summer, and none but open grate-fires are allowed. Insomnia never troubles him, for he falls asleep as soon as his head touches the pillow. His appetite and digestion are always good, and he has not lost a meal in ten years. What an example to the men who imagine it is hard work that is killing them in this career of unremitting industry!"

To return to "Screw" Smith, after the successful trial of his little vessel at sea in the autumn of 1837. He had many difficulties yet to contend with. There was, first, the difficulty of a new invention, and the fact that the paddle-boat had established itself in public estimation. The engineering and shipbuilding world were dead against him. They regarded the project of propelling a vessel by means of a screw as visionary and preposterous. There was also the official unwillingness to undertake anything novel, untried, and contrary to routine. There was the usual shaking of the head and the shrugging of the shoulders, as if the inventor were either a mere dreamer or a projector eager to lay his hands upon the public purse. The surveyor of the navy was opposed to the plan, because of the impossibility of making a vessel steer which was impelled from the stern. "Screw" Smith bided his time; he continued undaunted, and was determined to succeed. He laboured steadily onward, maintaining his own faith unshaken, and upholding the faith of the gentlemen who had become associated with him in the prosecution of the invention.

At the beginning of 1838 the Lords of the Admiralty requested Mr. Smith to allow his vessel to be tried under their inspection. Two trials were accordingly made, and they gave so much satisfaction that the adoption of the propeller for naval purposes was considered as a not improbable contingency. Before deciding finally upon its adoption, the Lords of the Admiralty were anxious to see an experiment made with a vessel of not less than 200 tons. Mr. Smith had not the means of accomplishing this by himself, but with the improved prospects of the invention, capitalists now came to his aid. One of the most effective and energetic of these was Mr. Henry Currie, banker; and, with the assistance of others, the "Ship Propeller Company" was formed, and proceeded to erect the test ship proposed by the Admiralty.

The result was the Archimedes, a wooden vessel of 237 tons burthen. She was designed by Mr. Pasco, laid down by Mr. Wimshurst in the spring of 1838, was launched on the 18th of October following, and made her first trip in May 1839. She was fitted with a screw of one turn placed in the dead wood, and propelled by a pair of engines of 80-horse power. The vessel was built under the persuasion that her performance would be considered satisfactory if a speed was attained of four or five knots an hour, where as her actual speed was nine and a half knots. The Lords of the Admiralty were invited to inspect the ship. At the second trial Sir Edward Parry, Sir William Symonds, Captain Basil Hall, and other distinguished persons were present.

The results were again satisfactory. The success of the Archimedes astonished the engineering world. Even the Surveyor of the Royal Navy found that the vessel could steer! The Lords of the Admiralty could no longer shut their eyes. But the invention could not at once be adopted. It must be tested by the best judges. The vessel was sent to Dover to be tried with the best packets between Dover and Calais. Mr. Lloyd, the chief engineer of the Navy, conducted the investigation, and reported most favourably as to the manner of her performance. Yet several years elapsed before the screw was introduced into the service.

In 1840 the Archimedes was placed at the disposal of Captain Chappell, of the Royal Navy, who, accompanied by Mr. Smith, visited every principal port in Great Britain. She was thus seen by shipowners, marine engineers, and shipbuilders in every part of the kingdom. They regarded her with wonder and admiration; yet the new mode of navigation was not speedily adopted. The paddle-wheel still held its own. The sentiment, if not the plant and capital, of the engineering world, were against the introduction of the screw. After the vessel had returned from her circumnavigation of Great Britain, she was sent to Oporto, and performed the voyage in sixty-eight and a half hours, then held to be the quickest voyage on record. She was then sent to the Texel at the request of the Dutch Government. She went through the North Holland Canal, visited Amsterdam, Antwerp, and other ports; and everywhere left the impression that the screw was an efficient and reliable power in the propulsion of vessels at sea.

Shipbuilders, however, continued to "fight shy" of the screw. The late Isambard Kingdon Brunel is entitled to the credit of having first directed the attention of shipbuilders to this important invention. He was himself a man of original views, free from bias, and always ready to strike out a fresh path in engineering works. He was building a large new iron steamer at Bristol, the Great Britain, for passenger traffic between England and America. He had intended to construct her as a paddle steamer; but hearing of the success of the Archimedes, he inspected the vessel, and was so satisfied with the performance of the screw that he recommended his directors to adopt this method for propelling the Great Britain. His advice was adopted, and the vessel was altered so as to adapt her for the reception of the screw. The vessel was found perfectly successful, and on her first voyage to London she attained the speed of ten knots an hour, though the wind and balance of tides were against her. A few other merchant ships were built and fitted with the screw; the Princess Royal at Newcastle in 1840, the Margaret and Senator at Hull, and the Great Northern at Londonderry, in 1841.

The Lords of the Admiralty made slow progress in adapting the screw for the Royal Navy. Sir William Symonds, the surveyor and principal designer of Her Majesty's ships, was opposed to all new projects. He hated steam power, and was utterly opposed to iron ships. He speaks of them in his journal as "monstrous."[7] So long as he remained in office everything was done in a perfunctory way. A small vessel named the Bee was built at Chatham in 1841, and fitted with both paddles and the screw for the purposes of experiment. In the same year the Rattier, the first screw vessel built for the navy, was laid down at Sheerness. Although of only 888 tons burthen, she was not launched until the spring of 1843. She was then fitted with the same kind of screw as the Archimedes, that is, a double-headed screw of half a convolution. Experiments went on for about three years, so as to determine the best proportions of the screw, and the proportions then ascertained have since been the principal guides of engineering practice.

The Rattler was at length tried in a water tournament with the paddle-steamer Alecto, and signally defeated her. Francis Pettit Smith, like Gulliver, may be said to have dragged the whole British fleet after him. Were the paddle our only means of propulsion, our whole naval force would be reduced to a nullity. Hostile gunners would wing a paddle-steamer as effectually as a sportsman wings a bird, and all the plating in the world would render such a ship a mere helpless log on the water.

The Admiralty could no longer defer the use of this important invention. Like all good things, it made its way slowly and by degrees. The royal naval authorities, who in 1833 backed the side paddles, have since adopted the screw in most of the ships-of-war. In all long sea-going voyages, also, the screw is now the favourite mode of propulsion. Screw ships of prodigious size are now built and launched in all the ship-building ports of Britain, and are sent out to navigate in every part of the world.

The introduction of iron as the material for shipbuilding has immensely advanced the interests of steam navigation, as it enables the builders to construct vessels of great size with the finest lines, so as to attain the highest rates of speed.

One might have supposed that Francis Pettit Smith would derive some substantial benefit from his invention, or at least that the Ship Propeller Company would distribute large dividends among their proprietors. Nothing of the kind. Smith spent his money, his labour, and his ingenuity in conferring a great public benefit without receiving any adequate reward; and the company, instead of distributing dividends, lost about 50,000L. in introducing this great invention; after which, in 1856, the patent-right expired. Three hundred and twenty-seven ships and vessels of all classes in the Royal Navy had then been fitted with the screw propeller, and a much larger number in the merchant service; but since that time the number of screw propellers constructed is to be counted by thousands.

In his comparatively impoverished condition it was found necessary to do something for the inventor. The Civil Engineers, with Robert Stephenson, M.P., in the chair, entertained him at a dinner and presented him with a handsome salver and claret jug. And that he might have something to put upon his salver and into his claret jug, a number of his friends and admirers subscribed over 2000L. as a testimonial. The Government appointed him Curator of the Patent Museum at South Kensington; the Queen granted him a pension on the Civil List for 200L. a year; he was raised to the honour of knighthood in 1871, and three years later he died.

Francis Pettit Smith was not a great inventor. He had, like many others, invented a screw propeller. But, while those others had given up the idea of prosecuting it to its completion, Smith stuck to his invention with determined tenacity, and never let it go until he had secured for it a complete triumph. As Mr. Stephenson observed at the engineer's meeting: "Mr. Smith had worked from a platform which might have been raised by others, as Watt had done, and as other great men had done; but he had made a stride in advance which was almost tantamount to a new invention. It was impossible to overrate the advantages which this and other countries had derived from his untiring and devoted patience in prosecuting the invention to a successful issue." Baron Charles Dupin compared the farmer Smith with the barber Arkwright: "He had the same perseverance and the same indomitable courage. These two moral qualities enabled him to triumph over every obstacle." This was the merit of "Screw" Smith—that he was determined to realize what his predecessors had dreamt of achieving; and he eventually accomplished his great purpose.

Footnotes for Chapter II.

[1] In the Transactions of the Institution of Naval Architects for 1860, it was pointed out that the general dimensions and form of bottom of this ship were very similar to the most famous line-of-battle ships built down to the end of last century, some of which were then in existence.

[2] According to the calculation of Mr. Chatfield, of Her Majesty's dockyard at Plymouth, in a paper read before the British Association in 1841 on shipbuilding.

[3] The phrase "wooden walls" is derived from the Greek. When the city of Athens was once in danger of being attacked and destroyed, the oracle of Delphi was consulted. The inhabitants were told that there was no safety for them but in their "wooden walls,"—that is their shipping. As they had then a powerful fleet, the oracle gave them rational advice, which had the effect of saving the Athenian people.

[4] An account of these is given by Bennet Woodcraft in his Sketch of the Origin and Progress of Steam Navigation, London, 1848.

[5] See Industrial Biography, pp. 183-197,

[6] The story is told in Scribner's Monthly Illustrated Magazine, for April 1879. Ericsson's modest bill was only $15,000 for two years' labour. He was put off from year to year, and at length the Government refused to pay the

amount. "The American Government," says the editor of Scribner, "will not appropriate the money to pay it, and that is all. It is said to be the nature of republics to be ungrateful; but must they also be dishonest?"

[7] Memoirs of the Life and Services of Rear-Admiral Sir William Symonds, Kt., p. 332.

CHAPTER III.[1]

JOHN HARRISON: INVENTOR OF THE MARINE CHRONOMETER.

"No man knows who invented the mariner's compass, or who first hollowed out a canoe from a log. The power to observe accurately the sun, moon, and planets, so as to fix a vessel's actual position when far out of sight of land, enabling long voyages to be safely made; the marvellous improvements in ship-building, which shortened passages by sailing vessels, and vastly reduced freights even before steam gave an independent force to the carrier—each and all were done by small advances, which together contributed to the general movement of mankind.... Each owes all to the others. The forgotten inventors live for ever in the usefulness of the work they have done and the progress they have striven for."—H. M. Hyndman.

One of the most extraordinary things connected with Applied Science is the method by which the Navigator is enabled to find the exact spot of sea on which his ship rides. There may be nothing but water and sky within his view; he may be in the midst of the ocean, or gradually nearing the land; the curvature of the globe baffles the search of his telescope; but if he have a correct chronometer, and can make an astronomical observation, he may readily ascertain his longitude, and know his approximate position—how far he is from home, as well as from his intended destination. He is even enabled, at some special place, to send down his grappling-irons into the sea, and pick up an electrical cable for examination and repair.

This is the result of a knowledge of Practical Astronomy. "Place an astronomer," says Mr. Newcomb, "on board a ship; blindfold him; carry him by any route to any ocean on the globe, whether under the tropics or in one of the frigid zones; land him on the wildest rock that can be found; remove his bandage, and give him a chronometer regulated to Greenwich or Washington time, a transit instrument with the proper appliances, and the necessary books and tables, and in a single clear night he can tell his position within a hundred yards by observations of the stars.

This, from a utilitarian point of view, is one of the most important operations of Practical Astronomy."[2]

The Marine Chronometer was the outcome of the crying want of the sixteenth century for an instrument that should assist the navigator to find his longitude on the pathless ocean. Spain was then the principal naval power; she was the most potent monarchy in Europe, and held half America under her sway. Philip III. offered 100,000 crowns for any discovery by means of which the longitude might be determined by a better method than by the log, which was found very defective. Holland next became a great naval power, and followed the example of Spain in offering 30,000 florins for a similar discovery. But though some efforts were made, nothing practical was done, principally through the defective state of astronomical instruments. England succeeded Spain and Holland as a naval power; and when Charles II. established the Greenwich Observatory, it was made a special point that Flamsteed, the Astronomer-Royal, should direct his best energies to the perfecting of a method for finding the longitude by astronomical observations. But though Flamsteed, together with Halley and Newton, made some progress, they were prevented from obtaining ultimate success by the want of efficient chronometers and the defective nature of astronomical instruments.

Nothing was done until the reign of Queen Anne, when a petition was presented to the Legislature on the 25th of May, 1714, by "several captains of Her Majesty's ships, merchants in London, and commanders of merchantmen, in behalf of themselves, and of all others concerned in the navigation of Great Britain," setting forth the importance of the accurate discovery of the longitude, and the inconvenience and danger to which ships were subjected from the want of some suitable method of discovering it. The petition was referred to a committee, which took evidence on the subject. It appears that Sir Isaac Newton, with his extraordinary sagacity, hit the mark in his report. "One is," he said, "by a watch to keep time exactly; but, by reason of the motion of a ship, and the variation of heat and cold, wet and dry, and the difference of gravity in different latitudes, such a watch hath not yet been made."

An Act was however passed in the Session of 1714, offering a very large public reward to inventors: 10,000L. to any one who should discover a method of determining the longitude to one degree of a great circle, or 60 geographical miles; 15,000L. if it determined the same to two-thirds of that distance, or 40 geographical miles; and 20,000L. if it determined the same to one-half of the same distance, or 30 geographical miles. Commissioners were appointed by the same Act, who were instructed that "one moiety or half part of such reward shall be due and paid when the said commissioners, or the major part of them, do agree that any such method extends to the security of ships within 80 geographical miles of the shore, which are places of the greatest danger; and the other moiety or half part when a ship, by the appointment of the said commissioners, or the major part of them, shall actually sail over the ocean, from Great Britain to any such port in the West Indies as those commissioners, or the major part of them, shall choose or nominate for the experiment, without losing the longitude beyond the limits before mentioned."

The terms of this offer indicate how great must have been the risk and inconvenience which it was desired to remedy. Indeed, it is almost inconceivable that a reward so great could be held out for a method which would merely afford security within eighty geographical miles!

This splendid reward for a method of discovering the longitude was offered to the world—to inventors and scientific men of all countries—without restriction of race, or nation, or language. As might naturally be expected, the prospect of obtaining it stimulated many ingenious men to make suggestions and contrive experiments; but for many years the successful construction of a marine time-keeper seemed almost hopeless. At length, to the surprise of every one, the prize was won by a village carpenter—a person of no school, or university, or college whatever.

Even so distinguished an artist and philosopher as Sir Christopher Wren was engaged, as late in his life as the year 1720, in attempting to solve this important problem. As has been observed, in the memoir of him contained in the 'Biographia Britannica,'[3] "This noble invention, like some others of the most useful ones to human life, seems to be reserved for the peculiar glory of an ordinary mechanic, who, by indefatigable industry, under the guidance of no ordinary sagacity, hath seemingly at last surmounted all difficulties, and brought it to a most unexpected degree of perfection." Where learning and science failed, natural genius seems to have triumphed.

The truth is, that the great mechanic, like the great poet, is born, not made; and John Harrison, the winner of the famous prize, was a born mechanic. He did not, however, accomplish his object without the exercise of the greatest skill, patience, and perseverance. His efforts were long, laborious, and sometimes apparently hopeless. Indeed, his life, so far as we can ascertain the facts, affords one of the finest examples of difficulties encountered and triumphantly overcome, and of undaunted perseverance eventually crowned by success, which is to be found in the whole range of biography.

No complete narrative of Harrison's career was ever written. Only a short notice of him appears in the 'Biographia Britannica,' published in 1766, during his lifetime'—the facts of which were obtained from himself. A few notices of him appear in the 'Annual Register,' also published during his lifetime. The final notice appeared in the volume published in 1777, the year after his death. No Life of him has since appeared. Had he been a destructive hero, and fought battles by land or sea, we should have had biographies of him without end. But he pursued a more peaceful and industrious course. His discovery conferred an incalculable advantage on navigation, and enabled innumerable lives to be saved at sea; it also added to the domains of science by its more exact measurement of time. But his memory has been suffered to pass silently away, without any record being left for the benefit and advantage of those who have succeeded him. The following memoir includes nearly all that is known of the life and labours of John Harrison.

He was born at Foulby, in the parish of Wragby, near Pontefract, Yorkshire, in March, 1693. His father, Henry Harrison, was carpenter and joiner to Sir Rowland Winn, owner of the Nostell

Priory estate. The present house was built by the baronet on the site of the ancient priory. Henry Harrison was a sort of retainer of the family, and long continued in their Service.

Little is known of the boy's education. It was certainly of a very inferior description. Like George Stephenson, Harrison always had a great difficulty in making himself understood, either by speech or writing. Indeed, every board-school boy now receives a better education than John Harrison did a hundred and eighty years ago. But education does not altogether come by reading and writing. The boy was possessed of vigorous natural abilities. He was especially attracted by every machine that moved upon wheels. The boy was 'father to the man.' When six years old, and lying sick of small-pox, a going watch was placed upon his pillow, which afforded him infinite delight.

When seven years old he was taken by his father to Barrow, near Barton-on-Humber, where Sir Rowland Winn had another residence and estate. Henry Harrison was still acting as the baronet's carpenter and joiner. In course of time young Harrison joined his father in the workshop, and proved of great use to him. His opportunities for acquiring knowledge were still very few, but he applied his powers of observation and his workmanship upon the things which were nearest him. He worked in wood, and to wood he first turned his attention.

He was still fond of machines going upon wheels. He had enjoyed the sight of the big watch going upon brass wheels when he was a boy; but, now that he was a workman in wood, he proposed to make an eight-day clock, with wheels of this material. He made the clock in 1713, when he was twenty years old,[4] so that he must have made diligent use of his opportunities. He had of course difficulties to encounter, and nothing can be accomplished without them; for it is difficulties that train the habits of application and perseverance. But he succeeded in making an effective clock, which counted the time with regularity. This clock is still in existence. It is to be seen at the Museum of Patents, South Kensington; and when we visited it a few months ago it was going, and still marking the moments as they passed. It is contained in a case about six feet high, with a glass front, showing a pendulum and two weights. Over the clock is the following inscription:

"This clock was made at Barrow, Lincolnshire, in the year 1715, by John Harrison, celebrated as the inventor of a nautical timepiece, or chronometer, which gained the reward of 20,000L., offered by the Board of Longitude, A.D. 1767.

"This clock strikes the hour, indicates the day of the month, and with one exception (the escapement) the wheels are entirely made of wood."

This, however, was only a beginning. Harrison proceeded to make better clocks; and then he found it necessary to introduce metal, which was more lasting. He made pivots of brass, which moved more conveniently in sockets of wood with the use of oil. He also caused the teeth of his wheels to run against cylindrical rollers of wood, fixed by brass pins, at a proper distance from the axis of the pinions; and thus to a considerable extent removed the inconveniences of friction.

In the meantime Harrison eagerly improved every incident from which he might derive further information. There was a clergyman who came every Sunday to the village to officiate in the neighbourhood; and having heard of the sedulous application of the young carpenter, he lent him a manuscript copy of Professor Saunderson's discourses. That blind professor had prepared several lectures on natural philosophy for the use of his students, though they were not intended for publication. Young Harrison now proceeded to copy them out, together with the diagrams. Sometimes, indeed, he spent the greater part of the night in writing or drawing.

As part of his business, he undertook to survey land, and to repair clocks and watches, besides carrying on his trade of a carpenter. He soon obtained a considerable knowledge of what had been done in clocks and watches, and was able to do not only what the best professional workers had done, but to strike out entirely new lights in the clock and watch-making business. He found out a method of diminishing friction by adding a joint to the pallets of the pendulum, whereby they were made to work in the nature of rollers of a large radius, without any sliding, as usual, upon the teeth of the wheel. He constructed a clock on the recoiling principle, which went perfectly, and never lost a minute within fourteen years. Sir Edmund Denison Beckett says that he invented this method in order to save himself the trouble of going so frequently to oil the escapement of a turret clock, of which he had charge; though there were other influences at work besides this.

But his most important invention, at this early period of his life, was his compensation pendulum. Every one knows that metals expand with heat and contract by cold. The pendulum of the clock therefore expanded in summer and contracted in winter, thereby interfering with the regular going of the clock. Huygens had by his cylindrical checks removed the great irregularity arising from the unequal lengths of the oscillations; but the pendulum was affected by the tossing of a ship at sea, and was also subject to a variation in weight, depending on the parallel of latitude. Graham, the well-known clock-maker, invented the mercurial compensation pendulum, consisting of a glass or iron jar filled with quicksilver and fixed to the end of the pendulum rod. When the rod was lengthened by heat, the quicksilver and the jar which contained it were simultaneously expanded and elevated, and the centre of oscillation was thus continued at the same distance from the point of suspension.

But the difficulty, to a certain extent, remained unconquered until Harrison took the matter in hand. He observed that all rods of metal do not alter their lengths equally by heat, or, on the contrary, become shorter by cold, but some more sensibly than others. After innumerable experiments Harrison at length composed a frame somewhat resembling a gridiron, in which the alternate bars were of steel and of brass, and so arranged that those which expanded the most were counteracted by those which expanded the least. By this means the pendulum contained the power of equalising its own action, and the centre of oscillation continued at the same absolute distance from the point of suspension through all the variations of heat and cold during the year.[5]

Thus by the year 1726, when he was only thirty-three years old, Harrison had furnished himself with two compensation clocks, in which all the irregularities to which these machines were subject, were either removed or so happily balanced, one metal against the other, that the two clocks kept time together in different parts of his house, without the variation of more than a single second in the month. One of them, indeed, which he kept by him for his own use, and constantly compared with a fixed star, did not vary so much as one whole minute during the ten years that he continued in the country after finishing the machine.[6]

Living, as he did, not far from the sea, Harrison next endeavoured to arrange his timekeeper for purposes of navigation.

He tried his clock in a vessel belonging to Barton-on-Humber; but his compensating pendulum could there be of comparatively little use; for it was liable to be tossed hither or thither by the sudden motions of the ship. He found it necessary, therefore, to mount a chronometer, or portable timekeeper, which might be taken from place to place, and subjected to the violent and irregular motion of a ship at sea, without affecting its rate of going. It was evident to him that the first mover must be changed from a weight and pendulum to a spring wound up and a compensating balance.

He now applied his genius in this direction. After pondering over the subject, he proceeded to London in 1728, and exhibited his drawings to Dr. Halley, then Astronomer-Royal. The Doctor referred him to Mr. George Graham, the distinguished horologer, inventor of the dead-beat escapement and the mercurial pendulum. After examining the drawings and holding some converse with Harrison, Graham perceived him to be a man of uncommon merit, and gave him every encouragement. He recommended him, however, to make his machine before again applying to the Board of Longitude.

Harrison returned home to Barrow to complete his task, and many years elapsed before he again appeared in London to present his first chronometer.

The remarkable success which Harrison had achieved in his compensating pendulum could not but urge him on to further experiments. He was no doubt to a certain extent influenced by the reward of 20,000L. which the English Government had offered for an instrument that should enable the longitude to be more accurately determined by navigators at sea than was then possible; and it was with the object of obtaining pecuniary assistance to assist him in completing his chronometer that Harrison had, in 1728, made his first visit to London to exhibit his drawings.

The Act of Parliament offering this superb reward was passed in 1714, fourteen years before, but no attempt had been made to claim it. It was right that England, then rapidly advancing to the first position as a commercial nation, should make every effort to render navigation less hazardous. Before correct chronometers were invented, or good lunar tables were prepared,[7] the ship, when fairly at sea, out of sight of land, and battling with the winds and tides, was in a measure lost. No method existed for accurately ascertaining the longitude. The ship might be out of its course for one or two hundred miles, for anything that the navigator knew; and only the wreck of his ship on some unknown coast told of the mistake that he had made in his reckoning.

It may here be mentioned that it was comparatively easy to determine the latitude of a ship at sea every day when the sun was visible. The latitude—that is, the distance of any spot from the equator and the pole—might be found by a simple observation with the sextant. The altitude of the sun at noon is found, and by a short calculation the position of the ship can be ascertained.

The sextant, which is the instrument universally used at sea, was gradually evolved from similar instruments used from the earliest times. The object of this instrument has always been to find the angular distance between two bodies—that is to say, the angle contained by two straight lines, drawn from those bodies to meet in the observer's eye. The simplest instrument of this kind may be well represented by a pair of compasses. If the hinge is held to the eye, one leg pointed to the distant horizon, and the other leg pointed to the sun, the position of the two legs will show the angular distance of the sun from the horizon at the moment of observation.

Until the end of the seventeenth century, the instrument used was of this simple kind. It was generally a large quadrant, with one or two bars moving on a hinge,—to all intents and purposes a huge pair of compasses. The direction of the sight was fixed by the use of a slit and a pointer, much as in the ordinary rifle. This instrument was vastly improved by the use of a telescope, which not only allowed fainter objects to be seen, but especially enabled the sight to be accurately directed to the object observed.

The instruments of the pre-telescopic age reached their glory in the hands of Tycho Brahe. He used magnificent instruments of the simple "pair of compasses" kind—circles, quadrants, and sextants. These were for the most part ponderous fixed instruments of little or no use for the purposes of navigation. But Tycho Brahe's sextant proved the forerunner of the modern instrument. The general structure is the same; but the vast improvement of the modern sextant is due, firstly, to the use of the reflecting mirror, and, secondly, to the use of the telescope for accurate sighting. These improvements were due to many scientific men—to William Gascoigne, who first used the telescope, about 1640; to Robert Hooke, who, in 1660, proposed to apply it to the quadrant; to Sir Isaac Newton, who designed a reflecting quadrant;[8] and to John Hadley, who introduced it. The modern sextant is merely a modification of Newton's or Badley's quadrant, and its present construction seems to be perfect.

It therefore became possible accurately to determine the position of a ship at sea as regarded its latitude. But it was quite different as regarded the longitude that is, the distance of any place from a given meridian, eastward or westward. In the case of longitude there is no fixed spot to which reference can be made. The rotation of the earth makes the existence of such a spot impossible. The question of longitude is purely a question of TIME. The circuit of the globe, east and west, is simply represented by twenty-four hours. Each place has its own time. It is very easy to determine the local time at any spot by observations made at that spot. But, as time is always changing, the knowledge of the local time gives no idea of the actual position; and still less of a moving object—say, of a ship at sea. But if, in any locality, we know the local time, and also the local time of some other locality at that moment—say, of the Observatory at Greenwich we can, by comparing the two local times, determine the difference of local times, or, what is the same thing, the difference of longitude between the two places. It was necessary therefore for the navigator to be in possession of a first-rate watch or chronometer, to enable him to determine accurately the position of his ship at sea, as respected the longitude.

Before the middle of the eighteenth century good watches were comparatively unknown. The navigator mainly relied, for his approximate longitude, upon his Dead Reckoning, without any observation of the heavenly bodies. He depended upon the accuracy of the course which he had steered by the compass, and the mensuration of the ship's velocity by an instrument called the Log, as well as by combining and rectifying all the allowances for drift, lee-way, and so on, according to the trim of the ship; but all of these were liable to much uncertainty, especially when the sea was in a boisterous condition. There was another and independent course which might have been adopted—that is, by observation of the moon, which is constantly moving amongst the stars from west to east. But until the middle of the eighteenth century good lunar tables were as much unknown as good watches.

Hence a method of ascertaining the longitude, with the same degree of accuracy which is attainable in respect of latitude, had for ages been the grand desideratum for men "who go down to the sea in ships." Mr. Macpherson, in his important work entitled 'The Annals of Commerce,' observes, "Since the year 1714, when Parliament offered a reward of 20,000L. for the best method of ascertaining the longitude at sea, many schemes have been devised, but all to little or no purpose, as going generally upon wrong principles, till that heaven-taught artist Mr. John Harrison arose;" and by him, as Mr. Macpherson goes on to say, the difficulty was conquered, having devoted to it "the assiduous studies of a long life."

The preamble of the Act of Parliament in question runs as follows: "Whereas it is well known by all that are acquainted with the art of navigation that nothing is so much wanted and desired at sea as the discovery of the longitude, for the safety and quickness of voyages, the preservation of ships and the lives of men," and so on. The Act proceeds to constitute certain persons commissioners for the discovery of the longitude, with power to receive and experiment upon proposals for that purpose, and to grant sums of money not exceeding 2000L. to aid in such experiments. It will be remembered from what has been above stated, that a reward of 10,000L. was to be given to the person who should contrive a method of determining the longitude within one degree of a great circle, or 60 geographical miles; 15,000L. within 40 geographical miles; and 20,000L. within 30 geographical miles.

It will, in these days, be scarcely believed that little more than a hundred and fifty years ago a prize of not less than ten thousand pounds should have been offered for a method of determining the longitude within sixty miles, and that double the amount should have been offered for a method of determining it within thirty miles! The amount of these rewards is sufficient proof of the fearful necessity for improvement which then existed in the methods of navigation. And yet, from the date of the passing of the Act in 1714 until the year 1736, when Harrison finished his first timepiece, nothing had been done towards ascertaining the longitude more accurately, even within the wide limits specified by the Act of Parliament. Although several schemes had been projected, none of them had proved successful, and the offered rewards therefore still remained unclaimed.

To return to Harrison. After reaching his home at Barrow, after his visit to London in 1728, he began his experiments for the construction of a marine chronometer. The task was one of no small difficulty. It was necessary to provide against irregularities arising from the motion of a ship at sea, and to obviate the effect of alternations of temperature in the machine itself, as well as the oil with which it was lubricated. A thousand obstacles presented themselves, but they were not enough to deter Harrison from grappling with the work he had set himself to perform.

Every one knows the beautiful machinery of a timepiece, and the perfect tools required to produce such a machine. Some of these tools Harrison procured in London, but the greater number he provided for himself; and many entirely new adaptations were required for his chronometer. As wood could no longer be exclusively employed, as in his first clock, he had to teach himself to work accurately and minutely in brass and other metals. Having been unable to obtain any assistance from the Board of Longitude, he was under the necessity, while carrying forward his experiments, of maintaining himself by still working at his trade of a carpenter and joiner. This will account for the very long period that elapsed before he could bring his chronometer to such a state as that it might be tried with any approach to certainty in its operations.

Harrison, besides his intentness and earnestness, was a cheerful and hopeful man. He had a fine taste for music, and organised and led the choir of the village church, which attained a high degree of perfection. He invented a curious monochord, which was not less accurate than his clocks in the mensuration of time. His ear was distressed by the ringing of bells out of tune, and he set himself to remedy them. At the parish church of Hull, for instance, the bells were harsh and disagreeable, and by the authority of the vicar and churchwardens he was allowed to put them into a state of exact tune, so that they proved entirely melodious.

But the great work of his life was his marine chronometer. He found it necessary, in the first place, to alter the first mover of his clock to a spring wound up, so that the regularity of the motion might be derived from the vibrations of balances, instead of those of a pendulum as in a standing clock. Mr. Folkes, President of the Royal Society, when presenting the gold medal to Harrison in 1749, thus describes the arrangement of his new machine. The details were obtained from Harrison himself, who was present. He had made use of two balances situated in the same plane, but vibrating in contrary directions, so that the one of these being either way assisted by the tossing of the ship, the other might constantly be just so much impeded by it at the same time. As the equality of the times of the vibrations of the balance of a pocket-watch is in a great measure owing to the spiral spring that lies under it, so the same was here performed by the like elasticity of four cylindrical springs or worms, applied near the upper and lower extremities of the two balances above described.

Then came in the question of compensation. Harrison's experience with the compensation pendulum of his clock now proved of service to him. He had proceeded to introduce a similar expedient in his proposed chronometer. As is well known to those who are acquainted with the nature of springs moved by balances, the stronger those springs are, the quicker the vibrations of the balances are performed, and vice versa; hence it follows that those springs, when braced by cold, or when relaxed by heat, must of necessity cause the timekeeper to go either faster or slower, unless some method could be found to remedy the inconvenience.

The method adopted by Harrison was his compensation balance, doubtless the backbone of his invention. His "thermometer kirb," he himself says, "is composed of two thin plates of brass and steel, riveted together in several places, which, by the greater expansion of brass than steel by heat and contraction by cold, becomes convex on the brass side in hot weather and convex on the steel side in cold weather; whence, one end being fixed, the other end obtains a motion corresponding with the changes of heat and cold, and the two pins at the end, between which the balance spring passes, and which it alternately touches as the spring bends and unbends itself, will shorten or lengthen the spring, as the change of heat or cold would otherwise require to be done by hand in the manner used for regulating a common watch." Although the method has since been improved upon by Leroy, Arnold, and Earnshaw, it was the beginning of all that has since been done in the perfection of marine chronometers. Indeed, it is amazing to think of the number of clever, skilful, and industrious men who have been engaged for many hundred years in the production of that exquisite fabric—so useful to everybody, whether scientific or otherwise, on land or sea the modern watch.

It is unnecessary here to mention in detail the particulars of Harrison's invention. These were published by himself in his 'Principles of Mr. Harrison's Timekeeper.' It may, however, be mentioned that he invented a method by which the chronometer might be kept going without losing any portion of time. This was during the process of winding up, which was done once in a day. While the mainspring was being wound up, a secondary one preserved the motion of the wheels and kept the machine going.

After seven years' labour, during which Harrison encountered and overcame numerous difficulties, he at last completed his first marine chronometer. He placed it in a sort of moveable frame, somewhat resembling what the sailors call a 'compass jumble,' but much more artificially and curiously made and arranged. In this state the chronometer was tried from time to time in a large barge on the river Humber, in rough as well as in smooth weather, and it was found to go perfectly, without losing a moment of time.

Such was the condition of Harrison's chronometer when he arrived with it in London in 1735, in order to apply to the commissioners appointed for providing a public reward for the discovery of the longitude at sea. He first showed it to several members of the Royal Society, who cordially approved of it. Five of the most prominent members—Dr. Bailey, Dr. Smith, Dr. Bradley, Mr. John Machin, and Mr. George Graham—furnished Harrison with a certificate, stating that the principles of his machine for measuring time promised a very great and sufficient degree of exactness. In consequence of this certificate, the machine, at the request of the inventor, and at the recommendation of the Lords of the Admiralty, was placed on board a man-of-war.

Sir Charles Wager, then first Lord of the Admiralty, wrote to the captain of the Centurion, stating that the instrument had been approved by mathematicians as the best that had been made for measuring time; and requesting his kind treatment of Mr. Harrison, who was to accompany it to Lisbon. Captain Proctor answered the First Lord from Spithead, dated May 17th, 1736, promising his attention to Harrison's comfort, but intimating his fear that he had attempted impossibilities. It is always so with a new thing. The first steam-engine, the first gaslight, the first locomotive, the first steamboat to America, the first electric telegraph, were all impossibilities!

This first chronometer behaved very well on the outward voyage in the Centurion. It was not affected by the roughest weather, or by the working of the ship through the rolling waves of the Bay of Biscay. It was brought back, with Harrison, in the Orford man-of-war, when its great utility was proved in a remarkable manner, although, from the voyage being nearly on a meridian, the risk of losing the longitude was comparatively small. Yet the following was the certificate of the captain of the ship, dated the 24th June, 1737: "When we made the land, the said land, according to my reckoning (and others), ought to have been the Start; but, before we knew what land it was, John Harrison declared to me and the rest of the ship's company that, according to his observations with his machine, it ought to be the Lizard—the which, indeed, it was found to be, his observation showing the ship to be more west than my reckoning, above one degree and twenty-six miles,"—that is, nearly ninety miles out of its course!

Six days later—that is, on the 30th June—the Board of Longitude met, when Harrison was present, and produced the chronometer with which he had made the voyage to Lisbon and back. The minute states: "Mr. John Harrison produced a new invented machine, in the nature of clockwork, whereby he proposes to keep time at sea with more exactness than by any other instrument or method hitherto contrived, in order to the discovery of the longitude at sea; and proposes to make another machine of smaller dimensions within the space of two years, whereby he will endeavour to correct some defects which he hath found in that already prepared, so as to render the same more perfect; which machine, when completed, he is desirous of having tried in one of His Majesty's ships that shall be bound to the West Indies; but at the same time represented that he should not be able, by reason of his necessitous circumstances, to go on and finish his said machine without assistance, and requested that he may be furnished with the sum of 500L., to put him in a capacity to perform the same, and to make a perfect experiment thereof."

The result of the meeting was that 500L. was ordered to be paid to Harrison, one moiety as soon as convenient, and the other when he has produced a certificate from the captain of one of His Majesty's ships that he has put the machine on board into the captain's possession. Mr. George Graham, who was consulted, urged that the Commissioners should grant Harrison at least 1000L., but they only awarded him half the sum, and at first only a moiety of the amount voted. At the recommendation of Lord Monson, who was present, Harrison accepted the 250L. as a help towards the heavy expenses which he had already incurred, and was again about to incur, in perfecting the invention. He was instructed to make his new chronometer of less dimensions, as the one exhibited was cumbersome and heavy, and occupied too much space on board.

He accordingly proceeded to make his second chronometer. It occupied a space of only about half the size of the first. He introduced several improvements. He lessened the number of the wheels, and thereby diminished friction. But the general arrangement remained the same. This second machine was finished in 1739. It was more simple in its arrangement, and less cumbrous in its dimensions. It answered even better than the first, and though it was not tried at sea its motions were sufficiently exact for finding the longitude within the nearest limits proposed by Act of Parliament.

Not satisfied with his two machines, Harrison proceeded to make a third. This was of an improved construction, and occupied still less space, the whole of the machine and its apparatus standing upon an area of only four square feet. It was in such forwardness in January, 1741, that it was exhibited before the Royal Society, and twelve of the most prominent members signed a certificate of "its great and excellent use, as well for determining the longitude at sea as for correcting the charts of the coasts." The testimonial concluded: "We do recommend Mr. Harrison to the favour of the Commissioners appointed by Act of Parliament as a person highly deserving of such further encouragement and assistance as they shall judge proper and sufficient to finish his third machine." The Commissioners granted him a further sum of 500L. Harrison was already reduced to necessitous circumstances by his continuous application to the improvement of the timekeepers. He had also got into debt, and required further assistance to enable him to proceed with their construction; but the Commissioners would only help him by driblets.

Although Harrison had promised that the third machine would be ready for trial on August 1, 1743, it was not finished for some years later. In June, 1746, we find him again appearing before the Board, asking for further assistance. While proceeding with his work he found it necessary to add a new spring, "having spent much time and thought in tempering them." Another 500L. was voted to enable him to pay his debts, to maintain himself and family, and to complete his chronometer.

Three years later he exhibited his third machine to the Royal Society, and on the 30th of November, 1749, he was awarded the Gold Medal for the year. In presenting it, Mr. Folkes, the President, said to Mr. Harrison, "I do here, by the authority and in the name of the Royal Society of London for the improving of natural knowledge, present you with this small but faithful token of their regard and esteem. I do, in their name congratulate you upon the successes you have already had, and I most sincerely wish that all your future trials may in every way prove answerable to these beginnings, and that the full accomplishment of your great undertaking may at last be crowned with all the reputation and advantage to yourself that your warmest wishes may suggest, and to which so many years so laudably and so diligently spent in the improvement of those talents which God Almighty has bestowed upon you, will so justly entitle your constant and unwearied perseverance."

Mr. Folkes, in his speech, spoke of Mr. Harrison as "one of the most modest persons he had ever known. In speaking," he continued, "of his own performances, he has assured me that, from the immense number of diligent and accurate experiments he has made, and from the severe tests to which he has in many ways put his instrument, he expects he shall be able with sufficient certainty, through all the greatest variety of seasons and the most irregular motions of the sea, to keep time constantly, without the variation of so much as three seconds in a week,—a degree of exactness that is astonishing and even stupendous, considering the immense number of difficulties, and those of very different sorts, which the author of these inventions must have had to encounter and struggle withal."

Although it is common enough now to make first-rate chronometers—sufficient to determine the longitude with almost perfect accuracy in every clime of the world—it was very different at that time, when Harrison was occupied with his laborious experiments. Although he considered his third machine to be the ne plus ultra of scientific mechanism, he nevertheless proceeded to construct a fourth timepiece, in the form of a pocket watch about five inches in diameter. He found the principles which he had adopted in his larger machines applied equally well in the smaller, and the performances of the last surpassed his utmost expectations. But in the meantime, as his third timekeeper was, in his opinion, sufficient to supply the requirements of the Board of Longitude as respected the highest reward offered, he applied to the Commissioners for leave to try that instrument on board a royal ship to some port in the West Indies, as directed by the statute of Queen Anne.

Though Harrison's third timekeeper was finished about the year 1758, it was not until March 12, 1761, that he received orders for his son William to proceed to Portsmouth, and go on board the Dorsetshire man-of-war, to proceed to Jamaica. But another tedious delay occurred. The ship was ordered elsewhere, and William Harrison, after remaining five months at Portsmouth, returned to London. By this time, John Harrison had finished his fourth timepiece—the small one, in the form of a watch. At length William Harrison set sail with this timekeeper from Portsmouth for Jamaica, on November 18th, 1761, in the Deptford man-of-war. The Deptford had forty-three ships in convoy, and arrived at Jamaica on the 19th of January, 1762, three days before the Beaver, another of His Majesty's ships-of-war, which had sailed from Portsmouth ten days before the Deptford, but had lost her reckoning and been deceived in her longitude, having trusted entirely to the log. Harrison's timepiece had corrected the log of the Deptford to the extent of three degrees of longitude, whilst several of the ships in the fleet lost as much as five degrees! This shows the haphazard way in which navigation was conducted previous to the invention of the marine chronometer.

When the Deptford arrived at Port Royal, Jamaica, the timekeeper was found to be only five and one tenth seconds in error; and during the voyage of four months, on its return to Portsmouth on March 26th, 1762, it was found (after allowing for the rate of gain or loss) to have erred only one minute fifty-four and a half seconds. In the latitude of Portsmouth this only amounted to eighteen geographical miles, whereas the Act had awarded that the prize should be given where the longitude was determined within the distance of thirty geographical miles. One would have thought that Harrison was now clearly entitled to his reward of 20,000L.

Not at all! The delays interposed by Government are long and tedious, and sometimes insufferable. Harrison had accomplished more than was needful to obtain the highest reward which the Board of Longitude had publicly offered. But they would not certify that he had won the prize. On the contrary, they started numerous objections, and continued for years to subject him to vexatious delays and disappointments. They pleaded that the previous determination of the longitude of Jamaica by astronomical observation was unsatisfactory; that there was no proof of the chronometer having maintained a uniform rate during the voyage; and on the 17th of August, 1762, they passed a resolution, stating that they "were of opinion that the experiments made of the watch had not been sufficient to determine the longitude at sea."

It was accordingly necessary for Harrison to petition Parliament on the subject. Three reigns had come and gone since the Act of Parliament offering the reward had been passed. Anne had died; George I. and George II. had reigned and died; and now, in the reign of George III.—thirty-five years after Harrison had begun his labours, and after he had constructed four several marine chronometers, each of which was entitled to win the full prize,—an Act of Parliament was passed enabling the inventor to obtain the sum of 5000L. as part of the reward. But the Commissioners still hesitated. They differed about the tempering of the springs. They must have another trial of the timekeeper, or anything with which to put off a settlement of the claim. Harrison was ready for any further number of trials; and in the meantime the Commissioners merely paid him a further sum on account.

Two more dreary years passed. Nothing was done in 1763 except a quantity of interminable talk at the Board of Commissioners. At length, on the 28th of March, 1764, Harrison's son again departed with the timekeeper on board the ship Tartar for Barbadoes. He returned in about four months, during which time the instrument enabled the longitude to be ascertained within ten miles, or one-third of the required geographical distance. Harrison memorialised the Commissioners again and again, in order that he might obtain the reward publicly offered by the Government.

At length the Commissioners could no longer conceal the truth. In September,1764, they virtually recognised Harrison's claim by paying him 1000L. on account; and, on the 9th of February,1765, they passed a resolution setting forth that they were "unanimously of opinion that the said timekeeper has kept its time with sufficient correctness, without losing its longitude in the voyage from Portsmouth to Barbadoes beyond the nearest limit required by the Act 12th of Queen Anne, but even considerably within the same." Yet they would not give Harrison the necessary certificate, though they were of opinion that he was entitled to be paid the full reward!

It is pleasant to contrast the generous conduct of the King of Sardinia with the procrastinating and illiberal spirit which Harrison met with in his own country. During the same year in which the above resolution was passed, the Sardinian minister ordered four of Harrison's timekeepers at the price of 1000L. each, at the special instance of the King of Sardinia "as an acknowledgement of Mr. Harrison's ingenuity, and as some recompense for the time spent by him for the general good of mankind." This grateful attention was all the more praiseworthy, as Sardinia could not in any way be regarded as a great maritime power.

Harrison was now becoming old and feeble. He had attained the age of seventy-four. He had spent forty long years in working out his invention. He was losing his eyesight, and could not afford to wait much longer. Still he had to wait.

"Full little knowest thou, who hast not tried,
What hell it is in suing long to bide;
To lose good days, that might be better spent;
To waste long nights in pensive discontent;
To spend to-day, to be put back to-morrow,
To feed on hope, to pine with fear and sorrow."

But Harrison had not lost his spirit. On May 30th, 1765, he addressed another remonstrance to the Board, containing much stronger language than he had yet used. "I cannot help thinking," he said, "that I am extremely ill-used by gentlemen from whom I might have expected a different treatment; for, if the Act of the 12th of Queen Anne be deficient, why have I so long been encouraged under it, in order to bring my invention to perfection? And, after the completion, why was my son sent twice to the West Indies? Had it been said to my son, when he received the last instruction, 'There will, in case you succeed, be a new Act on your return, in order to lay you under new restrictions, which were not thought of in the Act of the 12th of Queen Anne,'—I say, had this been the case, I might have expected some such treatment as that I now meet with.

"It must be owned that my case is very hard; but I hope I am the first, and for my country's sake I hope I shall be the last, to suffer by pinning my faith upon an English Act of Parliament. Had I received my just reward—for certainly it may be so called after forty years' close application of the talent which it has pleased God to give me—then my invention would have taken the course which all improvements in this world do; that is, I must have instructed workmen in its principles and execution, which I should have been glad of an opportunity of doing. But how widely different this is from what is now proposed, viz., for me to instruct people that I know nothing of, and such as may know nothing of mechanics; and, if I do not make them understand to their satisfaction, I may then have nothing!

"Hard fate indeed to me, but still harder to the world, which may be deprived of this my invention, which must be the case, except by my open and free manner in describing all the principles of it to gentlemen and noblemen who almost at all times have had free recourse to my instruments. And if any of these workmen have been so ingenious as to have got my invention, how far you may please to reward them for their piracy must be left for you to determine; and I must set myself down in old age, and thank God I can be more easy in that I have the conquest, and though I have no reward, than if I had come short of the matter and by some delusion had the reward!"

The Right Honourable the Earl of Egmont was in the chair of the Board of Longitude on the day when this letter was read—June 13, 1765. The Commissioners were somewhat startled by the tone which the inventor had taken. Indeed, they were rather angry. Mr. Harrison, who was in waiting, was called in. After some rather hot speaking, and after a proposal was made to Harrison which he said he would decline to accede to "so long as a drop of English blood remained in his body," he left the room. Matters were at length arranged. The Act of Parliament (5 Geo. III. cap. 20) awarded him, upon a full discovery of the principles of his time-keeper, the payment of such a sum, as with the 2500L. he had already received, would make one half of the reward; and the remaining half was to be paid when other chronometers had been made after his design, and their capabilities fully proved. He was also required to assign his four chronometers—one of which was styled a watch—to the use of the public.

Harrison at once proceeded to give full explanations of the principles of his chronometer to Dr. Maskelyne, and six other gentlemen, who had been appointed to receive them. He took his timekeeper to pieces in their presence, and deposited in their hands correct drawings of the same, with the parts, so that other skilful makers might construct similar chronometers on the same principles. Indeed, there was no difficulty in making them; after his explanations and drawings had been published. An exact copy of his last watch was made by the ingenious Mr. Kendal; and was used by Captain Cook in his three years' circumnavigation of the world, to his perfect satisfaction.

England had already inaugurated that series of scientific expeditions which were to prove so fruitful of results, and to raise her naval reputation to so great a height. In these expeditions, the officers, the sailors, and the scientific men, were constantly brought face to face with unforeseen difficulties and dangers, which brought forth their highest qualities as men. There was, however, some intermixture of narrowness in the minds of those who sent them forth. For instance, while Dr. Priestley was at Leeds, he was asked by Sir Joseph Banks to join Captain Cook's second expedition to the Southern Seas, as an astronomer. Priestley gave his assent, and made arrangements to set out. But some weeks later, Banks informed him that his appointment had been cancelled, as the Board of Longitude objected to his theology. Priestley's otherwise gentle nature was roused. "What I am, and what they are, in respect of religion," he wrote to Banks, in

December, 1771, "might easily have been known before the thing was proposed to me at all. Besides, I thought that this had been a business of philosophy, and not of divinity. If, however, this be the case, I shall hold the Board of Longitude in extreme contempt."

Captain Cook was appointed to the command of the Resolution, and Captain Wallis to the command of the Adventure, in November, 1771. They proceeded to equip the ships; and amongst the other instruments taken on board Captain Cook's ship, were two timekeepers, one made by Mr. Larcum Kendal, on Mr. Harrison's principles, and the other by Mr. John Arnold, on his own. The expedition left Deptford in April, 1772; and shortly afterwards sailed for the South Seas. "Mr. Kendal's watch" is the subject of frequent notices in Captain Cook's account. At the Cape of Good Hope, it is said to have "answered beyond all expectation." Further south, in the neighbourhood of Cape Circumcision, he says, "the use of the telescope is found difficult at first, but a little practice will make it familiar. By the assistance of the watch we shall be able to discover the greatest error this method of observing the longitude at sea is liable to." It was found that Harrison's watch was more correct than Arnold's, and when near Cape Palliser in New Zealand, Cook says, "this day at noon, when we attended the winding-up of the watches, the fusee of Mr. Arnold's would not turn round, so that after several unsuccessful trials we were obliged to let it go down." From this time, complete reliance was placed upon Harrison's chronometer. Some time later, Cook says, "I must here take notice that our longitude can never be erroneous while we have so good a guide as Mr. Kendal's watch." It may be observed, that at the beginning of the voyage, observations were made by the lunar tables; but these, being found unreliable, were eventually discontinued.

To return to Harrison. He continued to be worried by official opposition. His claims were still unsatisfied. His watch at home underwent many more trials. Dr. Maskelyne, the Royal Astronomer, was charged with being unfavourable to the success of chronometers, being deeply interested in finding the longitude by lunar tables; although this method is now almost entirely superseded by the chronometer. Harrison accordingly could not get the certificate of what was due to him under the Act of Parliament. Years passed before he could obtain the remaining amount of his reward. It was not until the year 1773, or forty-five years after the commencement of his experiments, that he succeeded in obtaining it. The following is an entry in the list of supplies granted by Parliament in that year: "June 14. To John Harrison, as a further reward and encouragement over and above the sums already received by him, for his invention of a timekeeper for ascertaining the longitude at sea, and his discovery of the principles upon which the same was constructed, 8570 pounds 0s. 0d."

John Harrison did not long survive the settlement of his claims; for he died on the 24th of March, 1776, at the age of eighty-three. He was buried at the south-west corner of Hampstead parish churchyard, where a tombstone was erected to his memory, and an inscription placed upon it commemorating his services. His wife survived him only a year; she died at seventy-two, and was buried in the same tomb. His son, William Harrison, F.R.S., a deputy-lientenant of the counties of Monmouth and Middlesex, died in 1815, at the ripe age of eighty-eight, and was also interred there. The tomb having stood for more than a century, became somewhat dilapidated;

when the Clock-makers' Company of the City of London took steps in 1879 to reconstruct it, and recut the inscriptions. An appropriate ceremony took place at the final uncovering of the tomb.

But perhaps the most interesting works connected with John Harrison and the great labour of his life, are the wooden clock at the South Kensington Museum, and the four chronometers made by him for the Government, which are still preserved at the Royal Observatory, Greenwich. The three early ones are of great weight, and can scarcely be moved without some bodily labour. But the fourth, the marine chronometer or watch, is of small dimensions, and is easily handled. It still possesses the power of going accurately; as does "Mr. Kendal's watch," which was made exactly after it. These will always prove the best memorials of this distinguished workman.

Before concluding this brief notice of the life and labours of John Harrison, it becomes me to thank most cordially Mr. Christie, Astronomer-Royal, for his kindness in exhibiting the various chronometers deposited at the Greenwich Observatory, and for his permission to inspect the minutes of the Board of Longitude, where the various interviews between the inventor and the commissioners, extending over many years, are faithfully but too procrastinatingly recorded. It may be finally said of John Harrison, that by his invention of the chronometer—the ever-sleepless and ever-trusty friend of the mariner—he conferred an incalculable benefit on science and navigation, and established his claim to be regarded as one of the greatest benefactors of mankind.

POstscript.—In addition to the information contained in this chapter, I have been recently informed by the Rev. Mr. Sankey, vicar of Wragby, that the family is quite extinct in the parish, except the wife of a plumber, who claims relationship with Harrison. The representative of the Winn family was created Lord St. Oswald in 1885. Harrison is not quite forgotten at Foulby. The house in which he was born was a low thatched cottage, with two rooms, one used as a living room, and the other as a sleeping room. The house was pulled down about forty years ago; but the entrance door, being of strong, hard wood, is still preserved. The vicar adds that young Harrison would lie out on the grass all night in summer time, studying the details of his wooden clock.

Footnotes to Chapter III.

[1] Originally published in Longmam's Magazine, but now rewritten and enlarged.

[2] Popular Astronomy. By Simon Newcomb, LL.D., Professor U.S. Naval Observatory.

[3] Biographia Britannica, vol. vi. part 2, p. 4375. This volume was published in 1766, before the final reward had been granted to Harrison.

[4] This clock is in the possession of Abraham Riley, of Bromley, near Leeds. He informs us that the clock is made of wood throughout, excepting the escapement and the dial, which are made of brass. It bears the mark of "John Harrison, 1713."

[5] Harrison's compensation pendulum was afterwards improved by Arnold, Earnshaw, and other English makers. Dent's prismatic balance is now considered the best.

[6] See Mr. Folkes's speech to the Royal Soc., 30th Nov., 1749.

[7] No trustworthy lunar tables existed at that time. It was not until the year 1753 that Tobias Mayer, a German, published the first lunar tables which could be relied upon. For this, the British Government afterwards awarded to Mayer's widow the sum of 5000L.

[8] Sir Isaac Newton gave his design to Edmund Halley, then Astronomer-Royal. Halley laid it on one side, and it was found among his papers after his death in 1742, twenty-five years after the death of Newton. A similar omission was made by Sir G. B. Airy, which led to the discovery of Neptune being attributed to Leverrier instead of to Adams.

CHAPTER IV.

JOHN LOMBE: INTRODUCER OF THE SILK INDUSTRY INTO ENGLAND.

"By Commerce are acquired the two things which wise men accompt of all others the most necessary to the well-being of a Commonwealth: That is to say, a general Industry of Mind and Hardiness of Body, which never fail to be accompanyed with Honour and Plenty. So that, questionless, when Commerce does not flourish, as well as other Professions, and when Particular Persons out of a habit of Laziness neglect at once the noblest way of employing their time and the fairest occasion for advancing their fortunes, that Kingdom, though otherwise never so glorious, wants something of being compleatly happy."—A Treatise touching the East India Trade (1695).

Industry puts an entirely new face upon the productions of nature. By labour man has subjugated the world, reduced it to his dominion, and clothed the earth with a new garment. The first rude plough that man thrust into the soil, the first rude axe of stone with which he felled the pine, the first rude canoe scooped by him from its trunk to cross the river and reach the greener fields beyond, were each the outcome of a human faculty which brought within his reach some physical comfort he had never enjoyed before.

Material things became subject to the influence of labour. From the clay of the ground, man manufactured the vessels which were to contain his food. Out of the fleecy covering of sheep, he made clothes for himself of many kinds; from the flax plant he drew its fibres, and made linen and cambric; from the hemp plant he made ropes and fishing nets; from the cotton pod he fabricated fustians, dimities, and calicoes. From the rags of these, or from weed and the shavings of wood, he made paper on which books and newspapers were printed. Lead was formed by him into printer's type, for the communication of knowledge without end.

But the most extraordinary changes of all were made in a heavy stone containing metal, dug out of the ground. With this, when smelted by wood or coal, and manipulated by experienced skill, iron was produced. From this extraordinary metal, the soul of every manufacture, and the mainspring perhaps of civilised society—arms, hammers, and axes were made; then knives, scissors, and needles; then machinery to hold and control the prodigious force of steam; and eventually railroads and locomotives, ironclads propelled by the screw, and iron and steel bridges miles in length.

The silk manufacture, though originating in the secretion of a tiny caterpillar, is perhaps equally extraordinary. Hundreds of thousands of pounds weight of this slender thread, no thicker than the filaments spun by a spider, give employment to millions of workers throughout the world. Silk, and the many textures wrought from this beautiful material, had long been known in the East; but the period cannot be fixed when man first divested the chrysalis of its dwelling, and discovered that the little yellow ball which adhered to the leaf of the mulberry tree, could be evolved into a slender filament, from which tissues of endless variety and beauty could be made. The Chinese were doubtless among the first who used the thread spun by the silkworm for the purposes of clothing. The manufacture went westward from China to India and Persia, and from thence to Europe. Alexander the Great brought home with him a store of rich silks from Persia Aristotle and Pliny give descriptions of the industrious little worm and its productions. Virgil is the first of the Roman writers who alludes to the production of silk in China; and the terms he employs show how little was then known about the article. It was introduced at Rome about the time of Julius Caesar, who displayed a profusion of silks in some of his magnificent theatrical spectacles. Silk was so valuable that it was then sold for an equal weight of gold. Indeed, a law was passed that no man should disgrace himself by wearing a silken garment. The Emperor Heliogabalus despised the law, and wore a dress composed wholly of silk. The example thus set was followed by wealthy citizens. A demand for silk from the East soon became general.

It was not until about the middle of the sixth century that two Persian monks, who had long resided in China, and made themselves acquainted with the mode of rearing the silkworm, succeeded in carrying the eggs of the insect to Constantinople. Under their direction they were hatched and fed. A sufficient number of butterflies were saved to propagate the race, and mulberry trees were planted to afford nourishment to the rising generations of caterpillars. Thus the industry was propagated. It spread into the Italian peninsula; and eventually manufactures of silk velvet, damask, and satin became established in Venice, Milan, Florence, Lucca, and other places.

Indeed, for several centuries the manufacture of silk in Europe was for the most part confined to Italy. The rearing of silkworms was of great importance in Modena, and yielded a considerable revenue to the State. The silk produced there was esteemed the best in Lombardy. Until the beginning of the sixteenth century, Bologna was the only city which possessed proper "throwing" mills, or the machinery requisite for twisting and preparing silken fibres for the weaver. Thousands of people were employed at Florence and Genoa about the same time in the silk manufacture. And at Venice it was held in such high esteem, that the business of a silk factory was considered a noble employment.[1]

It was long before the use of silk became general in England. "Silk," said an old writer, "does not immediately come hither from the Worm that spins and makes it, but passes many a Climate, travels many a Desert, employs many a Hand, loads many a Camel, and freights many a Ship before it arrives here; and when at last it comes, it is in return for other manufactures, or in exchange for our money."[2] It is said that the first pair of silk stockings was brought into England from Spain, and presented to Henry VIII. He had before worn hose of cloth. In the third year of Queen Elizabeth's reign, her tiring woman, Mrs. Montagu, presented her with a pair of black silk stockings as a New Year's gift; whereupon her Majesty asked if she could have any more, in which case she would wear no more cloth stockings. When James VI. of Scotland received the ambassadors sent to congratulate him upon his accession to the throne of Great Britain, he asked one of his lords to lend him his pair of silken hose, that he "might not appear a scrub before strangers." From these circumstances it will be observed how rare the wearing of silk was in England.

Shortly after becoming king, James I. endeavoured to establish the silk manufacture in England, as had already been successfully done in France. He gave every encouragement to the breeding of silkworms. He sent circular letters to all the counties of England, strongly recommending the inhabitants to plant mulberry trees. The trees were planted in many places, but the leaves did not ripen in sufficient time for the sustenance of the silkworms.

The same attempt was made at Inneshannon, near Bandon, in Ireland, by the Hugnenot refugees, but proved abortive. The climate proved too cold or damp for the rearing of silkworms with advantage. All that remains is "The Mulberry Field," which still retains its name. Nevertheless the Huguenots successfully established the silk manufacture at London and Dublin, obtaining the spun silk from abroad.

Down to the beginning of last century, the Italians were the principal producers of organzine or thrown silk; and for a long time they succeeded in keeping their art a secret. Although the silk manufacture, as we have seen, was introduced into this country by the Huguenot artizans, the price of thrown silk was so great that it interfered very considerably with its progress. Organzine was principally made within the dominions of Savoy, by means of a large and curious engine, the like of which did not exist elsewhere. The Italians, by the most severe laws, long preserved the mystery of the invention. The punishment prescribed by one of their laws to be inflicted upon anyone who discovered the secret, or attempted to carry it out of the Sardinian dominions, was death, with the forfeiture of all the goods the delinquent possessed; and the culprit was "to be afterwards painted on the outside of the prison walls, hanging to the gallows by one foot, with an inscription denoting the name and crime of the person, there to be continued for a perpetual mark of infamy."[3]

Nevertheless, a bold and ingenious man was found ready to brave all this danger in the endeavour to discover the secret. It may be remembered with what courage and determination the founder of the Foley family introduced the manufacture of nails into England. He went into the Danemora mine district, near Upsala in Sweden, fiddling his way among the miners; and after making two voyages, he at last wrested from them the secret of making nails, and introduced the new industry into the Staffordshire district.[4] The courage of John Lombe, who introduced the thrown-silk industry into England, was equally notable. He was a native of Norwich. Playfair, in his 'Family Antiquity' (vii. 312), says his name "may have been taken from the French Lolme, or de Lolme," as there were many persons of French and Flemish origin settled at Norwich towards the close of the sixteenth century; but there is no further information as to his special origin.

John Lombe's father, Henry Lombe, was a worsted weaver, and was twice married. By his first wife he had two sons, Thomas and Henry; and by his second, he had also two sons, Benjamin and John. At his death in 1695, he left his two brothers his "supervisors," or trustees, and directed them to educate his children in due time to some useful trade. Thomas, the eldest son, went to London. He was apprenticed to a trade, and succeeded in business, as we find him Sheriff of London and Middlesex in 1727, when in his forty-second year. He was also knighted in the same year, most probably on the accession of George II. to the throne.

John, the youngest son of the family, and half-brother of Thomas, was put an apprentice to a trade. In 1702, we find him at Derby, working as a mechanic with one Mr. Crotchet. This unfortunate gentleman started a small silk-mill at Derby, with the object of participating in the profits derived from the manufacture.

"The wear of silks," says Hutton, in his 'History of Derby,' "was the taste of the ladies, and the British merchant was obliged to apply to the Italian with ready money for the article at an exorbitant price." Crotchet did not succeed in his undertaking. "Three engines were found necessary for the process: he had but one. An untoward trade is a dreadful sink for money; and an imprudent tradesman is still more dreadful. We often see instances where a fortune would last a man much longer if he lived upon his capital, than if he sent it into trade. Crotchet soon became insolvent."

John Lombe, who had been a mechanic in Crotchet's silk mill, lost his situation accordingly. But he seems to have been possessed by an intense desire to ascertain the Italian method of silk-throwing. He could not learn it in England. There was no other method but going to Italy, getting into a silk mill, and learning the secret of the Italian art. He was a good mechanic and a clever draughtsman, besides being intelligent and fearless.

But he had not the necessary money wherewith to proceed to Italy.

His half-brother Thomas, however, was doing well in London, and was willing to help him with the requisite means. Accordingly, John set out for Italy, not long after the failure of Crotchet.

John Lombe succeeded in getting employment in a silk mill in Piedmont, where the art of silk-throwing was kept a secret. He was employed as a mechanic, and had thus an opportunity, in course of time, of becoming familiar with the operation of the engine. Hutton says that he bribed the workmen; but this would have been a dangerous step, and would probably have led to his expulsion, if not to his execution. Hutton had a great detestation of the first silk factory at Derby, where he was employed when a boy; and everything that he says about it must be taken cum grano salis. When the subject of renewing the patent was before Parliament in 1731, Mr. Perry, who supported the petition of Sir Thomas Lombe, said that "the art had been kept so secret in Piedmont, that no other nation could ever yet come at the invention, and that Sir Thomas and his brother resolved to make an attempt for the bringing of this invention into their own country. They knew that there would be great difficulty and danger in the undertaking, because the king of Sardinia had made it death for any man to discover this invention, or attempt to carry it out of his dominions. The petitioner's brother, however, resolved to venture his person for the benefit and advantage of his native country, and Sir Thomas was resolved to venture his money, and to furnish his brother with whatever sums should be necessary for executing so bold and so generous a design. His brother went accordingly over to Italy; and after a long stay and a great expense in that country, he found means to see this engine so often, and to pry into the nature of it so narrowly, that he made himself master of the whole invention and of all the different parts and motions belonging to it."

John Lombe was absent from England for several years. While occupied with his investigations and making his drawings, it is said that it began to be rumoured that the Englishman was prying into the secret of the silk mill, and that he had to fly for his life. However this may be, he got on board an English ship, and returned to England in safety. He brought two Italian workmen with him, accustomed to the secrets of the silk trade. He arrived in London in 1716, when, after conferring with his brother, a specification was prepared and a patent for the organzining of raw silk was taken out in 1718. The patent was granted for fourteen years.

In the meantime, John Lombe arranged with the Corporation of the town of Derby for taking a lease of the island or swamp on the river Derwent, at a ground rental of 8L. a year. The island, which was well situated for water-power, was 500 feet long and 52 feet wide. Arrangements were at once made for erecting a silk mill thereon, the first large factory in England. It was constructed entirely at the expense of his brother Thomas. While the building was in progress, John Lombe hired various rooms in Derby, and particularly the Town Hall, where he erected temporary engines turned by hand, and gave employment to a large number of poor people.

At length, after about three years' labour, the great silk mill was completed. It was founded upon huge piles of oak, from 16 to 20 feet long, driven into the swamp close to each other by an engine made for the purpose. The building was five stories high, contained eight large apartments, and had no fewer than 468 windows. The Lombes must have had great confidence in their speculation, as the building and the great engine for making the organzine silk, together with the other fittings, cost them about 30,000L.

One effect of the working of the mill was greatly to reduce the price of the thrown-silk, and to bring it below the cost of the Italian production. The King of Sardinia, having heard of the success of the Lombe's undertaking, prohibited the exportation of Piedmontese raw silk, which interrupted the course of their prosperity, until means were taken to find a renewed supply elsewhere.

And now comes the tragic part of the story, for which Mr. Hutton, the author of the 'History of Derby,' is responsible. As he worked in the silk mill when a boy, from 1730 to 1737, he doubtless heard it from the mill-hands, and there may be some truth in it, though mixed with a little romance. It is this:—

Hutton says of John Lombe, that he "had not pursued this lucrative commerce more than three or four years when the Italians, who felt the effects from their want of trade, determined his destruction, and hoped that that of his works would follow. An artful woman came over in the character of a friend, associated with the parties, and assisted in the business. She attempted to gain both the Italian workmen, and succeeded with one. By these two slow poison was supposed, and perhaps justly, to have been administered to John Lombe, who lingered two or three years in agony, and departed. The Italian ran away to his own country; and Madam was interrogated, but nothing transpired, except what strengthened suspicion." A strange story, if true.

Of the funeral, Hutton says:—"John Lombe's was the most superb ever known in Derby. A man of peaceable deportment, who had brought a beneficial manufactory into the place, employed the poor, and at advanced wages, could not fail meeting with respect, and his melancholy end with pity. Exclusive of the gentlemen who attended, all the people concerned in the works were invited. The procession marched in pairs, and extended the length of Full Street, the market-place, and Iron-gate; so that when the corpse entered All Saints, at St. Mary's Gate, the last couple left the house of the deceased, at the corner of Silk-mill Lane."

Thus John Lombe died and was buried at the early age of twenty-nine; and Thomas, the capitalist, continued the owner of the Derby silk mill. Hutton erroneously states that William succeeded, and that he shot himself. The Lombes had no brother of the name of William, and this part of Hutton's story is a romance.

The affairs of the Derby silk mill went on prosperously. Enough thrown silk was manufactured to supply the trade, and the weaving of silk became a thriving business. Indeed, English silk began to have a European reputation. In olden times it was said that "the stranger buys of the Englishman the case of the fox for a groat, and sells him the tail again for a shilling." But now the matter was reversed, and the saying was, "The Englishman buys silk of the stranger for twenty marks, and sells him the same again for one hundred pounds."

But the patent was about to expire. It had been granted for only fourteen years; and a long time had elapsed before the engine could be put in operation, and the organzine manufactured. It was the only engine in the kingdom. Joshua Gee, writing in 1731, says: "As we have but one Water Engine in the kingdom for throwing silk, if that should be destroyed by fire or any other accident, it would make the continuance of throwing fine silk very precarious; and it is very much to be doubted whether all the men now living in the kingdom could make another." Gee accordingly recommended that three or four more should be erected at the public expense, "according to the model of that at Derby."[5]

The patent expired in 1732. The year before, Sir Thomas Lombe, who had been by this time knighted, applied to Parliament for a prolongation of the patent. The reasons for his appeal were principally these: that before he could provide for the full supply of other silk proper for his purpose (the Italians having prohibited the exportation of raw silk), and before he could alter his engine, train up a sufficient number of workpeople, and bring the manufacture to perfection, almost all the fourteen years of his patent right would have expired. "Therefore," the petition to Parliament concluded, "as he has not hitherto received the intended benefit of the aforesaid patent, and in consideration of the extraordinary nature of this undertaking, the very great expense, hazard, and difficulty he has undergone, as well as the advantage he has thereby procured to the nation at his own expense, the said Sir Thomas Lombe humbly hopes that Parliament will grant him a further term for the sole making and using his engines, or such other recompense as in their wisdom shall seem meet."[6]

The petition was referred to a Committee. After consideration, they recommended the House of Commons to grant a further term of years to Sir Thomas Lombe. The advisers of the King, however, thought it better that the patent should not be renewed, but that the trade in silk should be thrown free to all. Accordingly the Chancellor of the Exchequer acquainted the House (14th March, 1731) that "His Majesty having been informed of the case of Sir Thomas Lombe, with respect to his engine for making organzine silk, had commanded him to acquaint this House, that His Majesty recommended to their consideration the making such provision for a recompense to Sir Thomas Lombe as they shall think proper."

The result was, that the sum of 14,000L. was voted and paid to Sir Thomas Lombe as "a reward for his eminent services done to the nation, in discovering with the greatest hazard and difficulty the capital Italian engines, and introducing and bringing the same to full perfection in this kingdom, at his own great expense."[7] The trade was accordingly thrown open. Silk mills were erected at Stockport and elsewhere; Hutton says that divers additional mills were erected in Derby; and a large and thriving trade was established. In 1850, the number employed in the silk manufacture exceeded a million persons. The old mill has recently become disused. Although supported by strong wooden supports, it showed signs of falling; and it was replaced by a larger mill, more suitable to modern requirements.

Footnotes for Chapter IV.

[1] "This was equally the case with two other trades;—those of glass-maker and druggist, which brought no contamination upon nobility in Venice. In a country where wealth was concentrated in the hands of the powerful, it was no doubt highly judicious thus to encourage its employment for objects of public advantage. A feeling, more or less powerful, has always existed in the minds of the high-born, against the employment of their time and wealth to purposes of commerce or manufactures. All trades, save only that of war, seem to have been held by them as in some sort degrading, and but little comporting with the dignity of aristocratic blood." Cabinet Cyclopedia—Silk Manufacture, p. 20.

[2] A Brief State of the Inland or Home Trade. (Pamphlet.) 1730.

[3] A Brief State of the Case relating to the Machine erected at Derby for making Italian Organzine Silk, which was discovered and brought into England with the utmost difficulty and hazard, and at the Sole Expense of Sir Thomas Lombe. House of Commons Paper, 28th January, 1731.

[4] Self-Help, p. 205.

[5] The Trade and Navigation of Great Britain considered, p. 94.

[6] The petition sets forth the merits of the machine at Derby for making Italian organzine silk—"a manufacture made out of fine raw silk, by reducing it to a hard twisted fine and even thread. This silk makes the warp, and is absolutely necessary to mix with and cover the Turkey and other coarser silks thrown here, which are used for Shute,—so that, without a constant supply of this fine Italian organzine silk, very little of the said Turkey or other silks could be used, nor could the silk weaving trade be carried on in England. This Italian organzine (or thrown) silk has in all times past been bought with our money, ready made (or worked) in Italy, for want of the art of making it here. Whereas now, by making it ourselves out of fine Italian raw silk, the nation saves near one-third part; and by what we make out of fine China raw silk, above one-half of the price we pay for it ready worked in Italy. The machine at Derby contains 97,746 wheels, movements, and individual parts (which work day and night), all which receive their motion from one large water-wheel, are governed by one regulator, and it employs about 300 persons to attend and supply it with work." In Bees Cyclopaedia (art. 'Silk Manufacture') there is a full description of the Piedmont throwing machine introduced to England by John Lombe, with a good plate of it.

[7] Sir Thomas Lombe died in 1738. He had two daughters. The first, Hannah, was married to Sir Robert Clifton, of Clifton, co. Notts; the second, Mary Turner, was married to James, 7th Earl of Lauderdale. In his will, he "recommends his wife, at the conclusion of the Darby concern," to distribute among his "principal servants or managers five or six hundred pounds."

CHAPTER V.

WILLIAM MURDOCK: HIS LIFE AND INVENTIONS.

"Justice exacts, that those by whom we are most benefited Should be most admired."—Dr. Johnson.

"The beginning of civilization is the discovery of some useful arts, by which men acquire property, comforts, or luxuries. The necessity or desire of preserving them leads to laws and social institutions... In reality, the origin as well as the progress and improvement of civil society is founded on mechanical and chemical inventions."—Sir Humphry Davy.

At the middle of last century, Scotland was a very poor country. It consisted mostly of mountain and moorland; and the little arable land it contained was badly cultivated. Agriculture was almost a lost art. "Except in a few instances," says a writer in the 'Farmers' Magazine' of 1803, "Scotland was little better than a barren waste." Cattle could with difficulty be kept alive; and the people in some parts of the country were often on the brink of starvation. The people were hopeless, miserable, and without spirit, like the Irish in their very worst times. After the wreck of the Darien expedition, there seemed to be neither skill, enterprise, nor money left in the country. What resources it contained were altogether undeveloped. There was little communication between one place and another, and such roads as existed were for the greater part of the year simply impassable.

There were various opinions as to the causes of this frightful state of things. Some thought it was the Union between England and Scotland; and Andrew Fletcher of Saltoun, "The Patriot," as he was called, urged its Repeal. In one of his publications, he endeavoured to show that about one-sixth of the population of Scotland was in a state of beggary—two hundred thousand vagabonds begging from door to door, or robbing and plundering people as poor as themselves.[1] Fletcher was accordingly as great a repealer as Daniel O'Connell in after times. But he could not get the people to combine. There were others who held a different opinion. They thought that something might be done by the people themselves to extricate the country from its miserable condition.

It still possessed some important elements of prosperity. The inhabitants of Scotland, though poor, were strong and able to work. The land, though cold and sterile, was capable of cultivation.

Accordingly, about the middle of last century, some important steps were taken to improve the general condition of things. A few public-spirited landowners led the way, and formed themselves into a society for carrying out improvements in agriculture. They granted long leases of farms as a stimulus to the most skilled and industrious, and found it to their interest to give the farmer a more permanent interest in his improvements than he had before enjoyed. Thus stimulated and encouraged, farming made rapid progress, especially in the Lothians; and the example spread into other districts. Banks were established for the storage of capital. Roads were improved, and communications increased between one part of the country and another. Hence trade and commerce arose, by reason of the facilities afforded for the interchange of traffic. The people, being fairly educated by the parish schools, were able to take advantage of these improvements. Sloth and idleness gradually disappeared, before the energy, activity, and industry which were called into life by the improved communications.

At the same time, active and powerful minds were occupied in extending the domain of knowledge. Black and Robison, of Glasgow, were the precursors of James Watt, whose invention of the condensing steam-engine was yet to produce a revolution in industrial operations, the like of which had never before been known. Watt had hit upon his great idea while experimenting with an old Newcomen model which belonged to the University of Glasgow. He was invited by Mr. Roebuck of Kinneil to make a working steam-engine for the purpose of pumping water from the coal-pits at Boroughstoness; but his progress was stopped by want of capital, as well as by want of experience. It was not until the brave and generous Matthew Boulton of Birmingham took up the machine, and backed Watt with his capital and his spirit, that Watt's enterprise had the remotest chance of success. Even after about twelve years' effort, the condensing steam-engine was only beginning, though half-heartedly, to be taken up and employed by colliery proprietors and cotton manufacturers. In developing its powers, and extending its uses, the great merits of William Murdock can never be forgotten. Watt stands first in its history, as the inventor; Boulton second, as its promoter and supporter; and Murdock third, as its developer and improver.

William Murdock was born on the 21st of August, 1754, at Bellow Mill, in the parish of Auchinleck, Ayrshire. His father, John, was a miller and millwright, as well as a farmer. His mother's maiden name was Bruce, and she used to boast of being descended from Robert Bruce, the deliverer of Scotland. The Murdocks, or Murdochs—for the name was spelt in either way—were numerous in the neighbourhood, and they were nearly all related to each other. They are supposed to have originally come into the district from Flanders, between which country and Scotland a considerable intercourse existed in the middle ages. Some of the Murdocks took a leading part in the construction of the abbeys and cathedrals of the North;[2] others were known as mechanics; but the greater number were farmers.

One of the best known members of the family was John Murdock, the poet Burns' first teacher. Burns went to his school at Alloway Mill, when he was six years old. There he learnt to read and write. When Murdock afterwards set up a school at Ayr, Burns, who was then fifteen, went to board with him. In a letter to a correspondent, Murdock said: "In 1773, Robert Burns came to board and lodge with me, for the purpose of revising his English grammar, that he might be better qualified to instruct his brothers and sisters at home. He was now with me day and night, in school, at all meals, and in all my walks." The pupil even shared the teacher's bed at night. Murdock lent the boy books, and helped the cultivation of his mind in many ways. Burns soon revised his English grammar, and learnt French, as well as a little Latin. Some time after, Murdock removed to London, and had the honour of teaching Talleyrand English during his residence as an emigrant in this country. He continued to have the greatest respect for his former pupil, whose poetry commemorated the beauties of his native district.

It may be mentioned that Bellow Mill is situated on the Bellow Water, near where it joins the river Lugar. One of Burns' finest songs begins:—

"Behind yon hills where Lugar flows."

That was the scene of William Murdock's boyhood. When a boy, he herded his father's cows along the banks of the Bellow; and as there were then no hedges, it was necessary to have some one to watch the cattle while grazing. The spot is still pointed out where the boy, in the intervals of his herding, hewed a square compartment out of the rock by the water side, and there burnt the splint coal found on the top of the Black Band ironstone. That was one of the undeveloped industries of Scotland; for the Scotch iron trade did not arrive at any considerable importance until about a century later.[3] The little cavern in which Murdock burnt the splint coal was provided with a fireplace and vent, all complete. It is possible that he may have there derived, from his experiments, the first idea of Gas as an illuminant.

Murdock is also said to have made a wooden horse, worked by mechanical power, which was the wonder of the district. On this mechanical horse he rode to the village of Cumnock, about two miles distant. His father's name is, however, associated with his own in the production of this machine. Old John Murdock had a reputation for intelligence and skill of no ordinary kind. When at Carron ironworks, in 1760, he had a pinton cast after a pattern which he had prepared. This is said to have been the first piece of iron-toothed gearing ever used in mill work. When I last saw it, the pinton was placed on the lawn in front of William Murdock's villa at Handsworth.

The young man helped his father in many ways. He worked in the mill, worked on the farm, and assisted in the preparation of mill machinery. In this way he obtained a considerable amount of general technical knowledge. He even designed and constructed bridges. He was employed to build a bridge over the river Nith, near Dumfries, and it stands there to this day, a solid and handsome structure. But he had an ambition to be something more than a country mason. He had heard a great deal about the inventions of James Watt; and he determined to try whether he could not get "a job" at the famous manufactory at Soho. He accordingly left his native place in the year 1777, in the twenty-third year of his age; and migrated southward. He left plenty of Murdocks behind him. There was a famous staff in the family, originally owned by William Murdock's grandfather, which bore the following inscription: "This staff I leave in pedigree to the oldest Murdock after me, in the parish of Auchenleck, 1745." This staff was lately held by Jean Murdock, daughter of the late William Murdock, joiner, cousin of the subject of this biography.

When William arrived at Soho in 1777 he called at the works to ask for employment. Watt was then in Cornwall, looking after his pumping engines; but he saw Boulton, who was usually accessible to callers of every rank. In answer to Murdock's enquiry whether he could have a job, Boulton replied that work was very slack with them, and that every place was filled up. During the brief conversation that took place, the blate young Scotchman, like most country lads in the presence of strangers, had some difficulty in knowing what to do with his hands, and unconsciously kept twirling his hat with them. Boulton's attention was attracted to the twirling hat, which seemed to be of a peculiar make. It was not a felt hat, nor a cloth hat, nor a glazed hat: but it seemed to be painted, and composed of some unusual material. "That seems to be a curious sort of hat," said Boulton, looking at it more closely; "what is it made of?" "Timmer, sir," said Murdock, modestly. "Timmer? Do you mean to say that it is made of wood?" "'Deed it is, sir." "And pray how was it made?" "I made it mysel, sir, in a bit laithey of my own contrivin'." "Indeed!"

Boulton looked at the young man again. He had risen a hundred degrees in his estimation. William was a good-looking fellow—tall, strong, and handsome—with an open intelligent countenance. Besides, he had been able to turn a hat for himself with a lathe of his own construction. This, of itself, was a sufficient proof that he was a mechanic of no mean skill. "Well!" said Boulton, at last, "I will enquire at the works, and see if there is anything we can set you to. Call again, my man."

"Thank you, sir," said Murdock, giving a final twirl to his hat.

Such was the beginning of William Murdock's connection with the firm of Boulton and Watt. When he called again he was put upon a trial job, and then, as he was found satisfactory, he was engaged for two years at 15s. a week when at home, 17s. when in the country, and 18s. when in London. Boulton's engagement of Murdock was amply justified by the result. Beginning as an ordinary mechanic, he applied himself diligently and conscientiously to his work, and gradually became trusted. More responsible duties were confided to him, and he strove to perform them to the best of his power. His industry, skilfulness, and steady sobriety, soon marked him for promotion, and he rose from grade to grade until he became Boulton and Watt's most trusted co-worker and adviser in all their mechanical undertakings of importance.

Watt himself had little confidence in Scotchmen as mechanics. He told Sir Waiter Scott that though many of them sought employment at his works, he could never get any of them to become first-rate workmen. They might be valuable as clerks and book-keepers, but they had an insuperable aversion to toiling long at any point of mechanism, so as to earn the highest wages paid to the workmen.[4] The reason no doubt was, that the working-people of Scotland were then only in course of education as practical mechanics; and now that they have had a century's discipline of work and technical training, the result is altogether different, as the engine-shops and shipbuilding-yards of the Clyde abundantly prove. Mechanical power and technical ability are the result of training, like many other things.

When Boulton engaged Murdock, as we have said, Watt was absent in Cornwall, looking after the pumping-engines which had been erected at several of the mines throughout that county. The partnership had only been in existence for three years, and Watt was still struggling with the difficulties which he had to surmount in getting the steam engine into practical use. His health was bad, and he was oppressed with frightful headaches. He was not the man to fight the selfishness of the Cornish adventurers. "A little more of this hurrying and vexation," he said, "will knock me up altogether." Boulton went to his help occasionally, and gave him hope and courage. And at length William Murdock, after he had acquired sufficient knowledge of the business, was able to undertake the principal management of the engines in Cornwall.

We find that in 1779, when he was only twenty-five years old, he was placed in this important position. When he went into Cornwall, he gave himself no rest until he had conquered the defects of the engines, and put them into thorough working order.

He devoted himself to his duties with a zeal and ability that completely won Watt's heart. When he had an important job in hand, he could scarcely sleep. One night at his lodgings at Redruth, the people were disturbed by a strange noise in his room. Several heavy blows were heard upon the floor. They started from their beds, rushed to Murdock's room, and found him standing in his shirt, heaving at the bedpost in his sleep, shouting "Now she goes, lads! now she goes!"

Murdock became a most popular man with the mine owners. He also became friendly with the Cornish workmen and engineers. Indeed, he fought his way to their affections. One day, some half-dozen of the mining captains came into his engine-room at Chacewater, and began to bully him. This he could not stand. He stript, selected the biggest, and put himself into a fighting attitude. They set to, and in a few minutes Murdock's powerful bones and muscles enabled him to achieve the victory. The other men, who had looked on fairly, without interfering, seeing the temper and vigour of the man they had bullied, made overtures of reconciliation. William was quite willing to be friendly. Accordingly they shook hands all round, and parted the best of friends. It is also said that Murdock afterwards fought a duel with Captain Trevethick, because of a quarrel between Watt and the mining engineer, in which Murdock conceived his master to have been unfairly and harshly treated.[5]

The uses of Watt's steam-engine began to be recognised as available for manufacturing purposes. It was then found necessary to invent some method by which continuous rotary motion should be secured, so as to turn round the moving machinery of mills. With this object Watt had invented his original wheel-engine. But no steps were taken to introduce it into practical use. At length he prepared a model, in which he made use of a crank connected with the working beam of the engine, so as to produce the necessary rotary motion.

There was no originality in this application. The crank was one of the most common of mechanical appliances. It was in daily use in every spinning wheel, and in every turner's and knife-grinder's foot-lathe. Watt did not take out a patent for the crank, not believing it to be patentable. But another person did so, thereby anticipating Watt in the application of the crank for producing rotary motion. He had therefore to employ some other method, and in the new contrivance he had the valuable help of William Murdock. Watt devised five different methods of securing rotary motion without using the crank, but eventually he adopted the "Sun-and-planet motion," the invention of Murdock. This had the singular property of going twice round for every stroke of the engine, and might be made to go round much oftener without additional machinery. The invention was patented in February, 1782, five Years after Murdock had entered the service of Boulton and Watt.

Murdock continued for many years busily occupied in superintending the Cornish steam-engines. We find him described by his employers as "flying from mine to mine," putting the engines to rights. If anything went wrong, he was immediately sent for. He was active, quick-sighted, shrewd, sober, and thoroughly trustworthy. Down to the year 1780, his wages were only a pound a week; but Boulton made him a present of ten guineas, to which the owners of the United Mines added another ten, in acknowledgment of the admirable manner in which he bad erected their new engine, the chairman of the company declaring that he was "the most obliging and industrious workman he had ever known." That he secured the admiration of the Cornish engineers may be obvious from the fact of Mr. Boaze having invited him to join in an engineering partnership; but Murdock remained loyal to the Birmingham firm, and in due time he had his reward.

He continued to be the "right hand man" of the concern in Cornwall. Boulton wrote to Watt, towards the end of 1782: "Murdock hath been indefatigable ever since he began. He has scarcely been in bed or taken necessary food. After slaving night and day on Thursday and Friday, a letter came from Wheal Virgin that he must go instantly to set their engine to work, or they would let out the fire. He went and set the engine to work; it worked well for the five or six hours he remained. He left it, and returned to the Consolidated Mines about eleven at night, and was employed about the engines till four this morning, and then went to bed. I found him at ten this morning in Poldice Cistern, seeking for pins and castors that had jumped out, when I insisted on his going home to bed."

On one occasion, when an engine superintended by Murdock stopped through some accident, the water rose in the mine, and the workmen were "drowned out." Upon this occurring, the miners went "roaring at him" for throwing them out of work, and threatened to tear him to pieces. Nothing daunted, he went through the midst of the men, repaired the invalided engine, and started it afresh.

When he came out of the engine-house, the miners cheered him vociferously and insisted upon carrying him home upon their shoulders in triumph!

Steam was now asserting its power everywhere. It was pumping water from the mines in Cornwall and driving the mills of the manufacturers in Lancashire. Speculative mechanics began to consider whether it might not be employed as a means of land locomotion. The comprehensive mind of Sir Isaac Newton had long before, in his 'Explanation of the Newtonian Philosophy,' thrown out the idea of employing steam for this purpose; but no practical experiment was made. Benjamin Franklin, while agent in London for the United Provinces of America, had a correspondence with Matthew Boulton, of Birmingham, and Dr. Darwin, of Lichfield, on the same subject. Boulton sent a model of a fire-engine to London for Franklin's inspection; but Franklin was too much occupied at the time by grave political questions to pursue the subject further. Erasmus Darwin's speculative mind was inflamed by the idea of a "fiery chariot," and he urged his friend Boulton to prosecute the contrivance of the necessary steam machinery.[6]

Other minds were at work. Watt, when only twenty-three years old, at the instigation of his friend Robison, made a model locomotive, provided with two cylinders of tin plate; but the project was laid aside, and was never again taken up by the inventor. Yet, in his patent of 1784, Watt included an arrangement by means of which steam-power might be employed for the purposes of locomotion. But no further model of the contrivance was made.

Meanwhile, Cugnot, of Paris, had already made a road engine worked by steam power. It was first tried at the Arsenal in 1769; and, being set in motion, it ran against a stone wall in its way and threw it down. The engine was afterwards tried in the streets of Paris. In one of the experiments it fell over with a crash, and was thenceforward locked up in the Arsenal to prevent its doing further mischief. This first locomotive is now to be seen at the Conservatoire des Arts et Metiers at Paris.

Murdock had doubtless heard of Watt's original speculations, and proceeded, while at Redruth, during his leisure hours, to construct a model locomotive after a design of his own. This model was of small dimensions, standing little more than a foot and a half high, though it was sufficiently large to demonstrate the soundness of the principle on which it was constructed. It was supported on three wheels, and carried a small copper boiler, heated by a spirit lamp, with a flue passing obliquely through it. The cylinder, of 3/4 inch diameter and 2-inch stroke, was fixed in the top of the boiler, the piston-rod being connected with the vibratory beam attached to the connecting-rod which worked the crank of the driving-wheel. This little engine worked by the expansive force of steam only, which was discharged into the atmosphere after it had done its work of alternately raising and depressing the piston in the cylinder.

Mr. Murdock's son, while living at Handsworth, informed the present writer that this model was invented and constructed in 1781; but, after perusing the correspondence of Boulton and Watt, we infer that it was not ready for trial until 1784. The first experiment was made in Murdock's own house at Redruth, when the little engine successfully hauled a model waggon round the room,—the single wheel, placed in front of the engine and working in a swivel frame, enabling it to run round in a circle.

Another experiment was made out of doors, on which occasion, small though the engine was, it fairly outran the speed of its inventor. One night, after returning from his duties at the mine at Redruth, Murdock went with his model locomotive to the avenue leading to the church, about a mile from the town. The walk was narrow, straight, and level. Having lit the lamp, the water soon boiled, and off started the engine with the inventor after it. Shortly after he heard distant shouts of terror. It was too dark to perceive objects, but he found, on following up the machine, that the cries had proceeded from the worthy vicar, who, while going along the walk, had met the hissing and fiery little monster, which he declared he took to be the Evil One in propria persona!

When Watt was informed of Murdock's experiments, he feared that they might interfere with his regular duties, and advised their discontinuance. Should Murdock still resolve to continue them, Watt urged his partner Boulton, then in Cornwall, that, rather than lose Murdock's services, they should advance him 100L.; and, if he succeeded within a year in making an engine capable of drawing a post-chaise carrying two passengers and the driver, at the rate of four miles an hour, that a locomotive engine business should be established, with Murdock as a partner. The arrangement, however, never proceeded any further. Perhaps a different attraction withdrew Murdock from his locomotive experiments. He was then paying attention to a young lady, the daughter of Captain Painter; and in 1785 he married her, and brought her home to his house in Cross Street, Redruth.

In the following year,—September, 1786—Watt says, in a letter to Boulton, "I have still the same opinion concerning the steam carriage, but, to prevent more fruitless argument about it, I have one of some size under hand. In the meantime, I wish William could be brought to do as we do, to mind the business in hand, and let such as Symington and Sadler throw away their time and money in hunting shadows." In a subsequent letter Watt expressed his gratification at finding "that William applies to his business." From that time forward, Murdock as well as Watt, dropped all further speculation on the subject, and left it to others to work out the problem of the locomotive engine. Murdock's model remained but a curious toy, which he took pleasure in exhibiting to his intimate friends; and, though he long continued to speculate about road locomotion, and was persuaded of its practicability, he abstained from embodying his ideas of the necessary engine in any complete working form.

Murdock nevertheless continued inventing, for the man who is given to invent, and who possesses the gift of insight, cannot rest. He lived in the midst of inventors. Watt and Boulton were constantly suggesting new things, and Murdock became possessed by the same spirit. In 1791 he took out his first patent. It was for a method of preserving ships' bottoms from foulness by the use of a certain kind of chemical paint. Mr. Murdock's grandson informs us that it was recently re-patented and was the cause of a lawsuit, and that Hislop's patent for revivifying gas-lime would have been an infringement, if it had not expired.

Murdock is still better known by his invention of gas for lighting purposes. Several independent inquirers into the constituents of Newcastle coal had arrived at the conclusion that nearly one-third of the substance was driven off in vapour by the application of heat, and that the vapour so driven off was inflammable. But no suggestion had been made to apply this vapour for lighting purposes until Murdock took the matter in hand. Mr. M. S. Pearse has sent us the following interesting reminiscence: "Some time since, when in the West of Cornwall, I was anxious to find out whether any one remembered Murdock. I discovered one of the most respectable and intelligent men in Camborne, Mr. William Symons, who not only distinctly remembered Murdock, but had actually been present on one of the first occasions when gas was used. Murdock, he says, was very fond of children, and not unfrequently took them into his workshop to show them what he was doing. Hence it happened that on one occasion this gentleman, then a boy of seven or eight, was standing outside Murdock's door with some other boys, trying to catch sight of some special mystery inside, for Dr. Boaze, the chief doctor of the place, and Murdock had been busy all the afternoon. Murdock came out, and asked my informant to run down to a shop near by for a thimble. On returning with the thimble, the boy pretended to have lost it, and, whilst searching in every pocket, he managed to slip inside the door of the workshop, and then produced the thimble. He found Dr. Boaze and Murdock with a kettle filled with coal. The gas issuing from it had been burnt in a large metal case, such as was used for blasting purposes. Now, however, they had applied a much smaller tube, and at the end of it fastened the thimble, through the small perforations made in which they burned a continuous jet for some time."[7]

After numerous experiments, Murdock had his house in Cross Street fitted up in 1792 for being lit by gas. The coal was subjected to heat in an iron retort, and the gas was conveyed in pipes to the offices and the different rooms of the house, where it was burned at proper apertures or burners.[8] Portions of the gas were also confined in portable vessels of tinned iron, from which it was burned when required, thus forming a moveable gas-light. Murdock had a gas lantern in regular use, for the purpose of lighting himself home at night across the moors, from the mines where he was working, to his home at Redruth. This lantern was formed by filling a bladder with gas and fixing a jet to the mouthpiece at the bottom of a glass lantern, with the bladder hanging underneath.

Having satisfied himself as to the superior economy of coal gas, as compared with oils and tallow, for the purposes of artificial illumination, Murdock mentioned the subject to Mr. James Watt, jun., during a brief visit to Soho in 1794, and urged the propriety of taking out a patent. Watt was, however, indifferent to taking out any further patents, being still engaged in contesting with the Cornish mine-owners his father's rights to the user of the condensing steam-engine. Nothing definite was done at the time. Murdock returned to Cornwall and continued his experiments. At the end of the same year he exhibited to Mr. Phillips and others, at the Polgooth mine, his apparatus for extracting gases from coal and other substances, showed it in use, lit the gas which issued from the burner, and showed its "strong and beautiful light." He afterwards exhibited the same apparatus to Tregelles and others at the Neath Abbey Company's ironworks in Glamorganshire.

Murdock returned to Soho in 1798, to take up his permanent residence in the neighbourhood. When the mine owners heard of his intention to leave Cornwall, they combined in offering him a handsome salary provided he would remain in the county; but his attachment to his friends at Soho would not allow him to comply with their request. He again urged the firm of Boulton and Watt to take out a patent for the use of gas for lighting purposes. But being still embroiled in their tedious and costly lawsuit, they were naturally averse to risk connection with any other patent. Watt the younger, with whom Murdock communicated on the subject, was aware that the current of gas obtained from the distillation of coal in Lord Dundonald's tar-ovens had been occasionally set fire to, and also that Bishop Watson and others had burned gas from coal, after conducting it through tubes, or after it had issued from the retort. Mr. Watt was, however, quite satisfied that Murdock was the first person who had suggested its economical application for public and private uses.

But he was not clear, after the legal difficulties which had been raised as to his father's patent rights, that it would be safe to risk a further patent for gas.

Mr. Murdock's suggestion, accordingly, was not acted upon. But he went on inventing in other directions. He thenceforward devoted himself entirely to mechanical pursuits. Mr. Buckle has said of him:—"The rising sun often found him, after a night spent in incessant labour, still at the anvil or turning-lathe; for with his own hands he would make such articles as he would not intrust to unskilful ones." In 1799 he took out a patent (No. 2340), embodying some very important inventions. First, it included the endless screw working into a toothed-wheel, for boring steam-cylinders, which is still in use. Second, the casting of a steam-jacket in one cylinder, instead of being made in separate segments bolted together with caulked joints, as was previously done. Third, the new double-D slide-valve, by which the construction and working of the steam-engine was simplified, and the loss of steam saved, as well as the cylindrical valve for the same purpose. And fourth, improved rotary engines. One of the latter was set to drive the machines in his private workshop, and continued in nearly constant work and in perfect use for about thirty years.

In 1801, Murdock sent his two sons William and John to the Ayr Academy, for the benefit of Scotch education. In the summer-time they spent their vacation at Bellow Mill, which their grandfather still continued to occupy. They fished in the river, and "caught a good many trout." The boys corresponded regularly with their father at Birmingham. In 1804, they seem to have been in a state of great excitement about the expected landing of the French in Scotland. The volunteers of Ayr amounted to 300 men, the cavalry to 150, and the riflemen to 50. "The riflemen," says John, "go to the seashore every Saturday to shoot at a target. They stand at 70 paces distant, and out of 100 shots they often put in 60 bullets!" William says, "Great preparations are still making for the reception of the French. Several thousand of pikes are carried through the town every week; and all the volunteers and riflemen have received orders to march at a moment's warning." The alarm, however, passed away. At the end of 1804, the two boys received prizes; William got one in arithmetic and another in the Rector's composition class; and John also obtained two, one in the mathematical class, and the other in French.

To return to the application of gas for lighting purposes. In 1801, a plan was proposed by a M. Le Blond for lighting a part of the streets of Paris with gas. Murdock actively resumed his experiments; and on the occasion of the Peace of Amiens in March, 1802, he made the first public exhibition of his invention. The whole of the works at Soho were brilliantly illuminated with gas.

The sight was received with immense enthusiasm. There could now be no doubt as to the enormous advantages of this method of producing artificial light, compared with that from oil or tallow. In the following year the manufacture of gas-making apparatus was added to the other branches of Boulton and Watts' business, with which Murdock was now associated,—and as much as from 4000L. to 5000L. of capital were invested in the new works. The new method of lighting speedily became popular amongst manufacturers, from its superior safety, cheapness, and illuminating power. The mills of Phillips and Lee of Manchester were fitted up in 1805; and those of Burley and Kennedy, also of Manchester, and of Messrs. Gott, of Leeds, in subsequent years.

Though Murdock had made the uses of gas-lighting perfectly clear, it was some time before it was proposed to light the streets by the new method. The idea was ridiculed by Sir Humphry Davy, who asked one of the projectors if he intended to take the dome of St. Paul's for a gasometer! Sir Waiter Scott made many clever jokes about those who proposed to "send light through the streets in pipes;" and even Wollaston, a well known man of science, declared that they "might as well attempt to light London with a slice from the moon." It has been so with all new projects—with the steamboat, the locomotive, and the electric telegraph. As John Wilkinson said of the first vessel of iron which he introduced, "it will be only a nine days' wonder, and afterwards a Columbus's egg."

On the 25th of February, 1808, Murdock read a paper before the Royal Society "On the Application of Gas from Coal to economical purposes." He gave a history of the origin and progress of his experiments, down to the time when he had satisfactorily lit up the premises of Phillips and Lee at Manchester. The paper was modest and unassuming, like everything he did.

It concluded:—"I believe I may, without presuming too much, claim both the first idea of applying, and the first application of this gas to economical purposes."[9] The Royal Society awarded Murdock their large Rumford Gold Medal for his communication.

In the following year a German named Wintzer, or Winsor, appeared as the promotor of a scheme for obtaining a royal charter with extensive privileges, and applied for powers to form a joint-stock company to light part of London and Westminster with gas. Winsor claimed for his method of gas manufacture that it was more efficacious and profitable than any then known or practised. The profits, indeed, were to be prodigious. Winsor made an elaborate calculation in his pamphlet entitled 'The New Patriotic Imperial and National Light and Heat Company,' from which it appeared that the net annual profits "agreeable to the official experiments" would amount to over two hundred and twenty-nine millions of pounds!—and that, giving over nine-tenths of that sum towards the redemption of the National Debt, there would still remain a total profit of 570L. to be paid to the subscribers for every 5L. of deposit! Winsor took out a patent for the invention, and the company, of which he was a member, proceeded to Parliament for an Act.

Boulton and Watt petitioned against the Bill, and James Watt, junior, gave evidence on the subject. Henry Brougham, who was the counsel for the petitioners, made great fun of Winsor's absurd speculations,[10] and the Bill was thrown out.

In the following year the London and Westminster Chartered Gas Light and Coke Company succeeded in obtaining their Act. They were not very successful at first. Many prejudices existed against the employment of the new light. It was popularly supposed that the gas was carried along the pipes on fire, and that the pipes must necessarily be intensely hot. When it was proposed to light the House of Commons with gas, the architect insisted on the pipes being placed several inches from the walls, for fear of fire; and, after the pipes had been fixed, the members might be seen applying their gloved hands to them to ascertain their temperature, and afterwards expressing the greatest surprise on finding that they were as cool as the adjoining walls.

The Gas Company was on the point of dissolution when Mr. Samuel Clegg came to their aid. Clegg had been a pupil of Murdock's, at Soho. He knew all the arrangements which Murdock had invented. He had assisted in fitting up the gas machinery at the mills of Phillips & Lee, Manchester, as well as at Lodge's Mill, Sowerby Bridge, near Halifax. He was afterwards employed to fix the apparatus at the Catholic College of Stoneyhurst, in Lancashire, at the manufactory of Mr. Harris at Coventry, and at other places. In 1813 the London and Westminster Gas Company secured the services of Mr. Clegg, and from that time forwards their career was one of prosperity. In 1814 Westminster Bridge was first lighted with gas, and shortly after the streets of St. Margaret's, Westminster. Crowds of people followed the lamplighter on his rounds to watch the sudden effect of his flame applied to the invisible stream of gas which issued from the burner. The lamplighters became so disgusted with the new light that they struck work, and Clegg himself had for a time to act as lamplighter.

The advantages of the new light, however, soon became generally recognised, and gas companies were established in most of the large towns. Glasgow was lit up by gas in 1817, and Liverpool and Dublin in the following year. Had Murdock in the first instance taken out a patent for his invention, it could not fail to have proved exceedingly remunerative to him; but he derived no advantage from the extended use of the new system of lighting except the honour of having invented it.[11] He left the benefits of his invention to the public, and returned to his labours at Soho, which more than ever completely engrossed him.

Murdock now became completely identified with the firm of Boulton & Watt. He assigned to them his patent for the slide-valve, the rotary engine, and other inventions "for a good and valuable consideration." Indeed his able management was almost indispensable to the continued success of the Soho foundry. Mr. Nasmyth, when visiting the works about thirty years after Murdock had taken their complete management in hand, recalled to mind the valuable services of that truly admirable yet modest mechanic. He observed the admirable system, which he had invented, of transmitting power from one central engine to other small vacuum engines attached to the several machines which they were employed to work. "This vacuum method," he says, "of transmitting power dates from the time of Papin; but it remained a dead contrivance for about a century until it received the masterly touch of Murdock."

"The sight which I obtained" (Mr. Nasmyth proceeds) "of the vast series of workshops of that celebrated establishment, fitted with evidences of the presence and results of such master minds in design and execution, and the special machine tools which I believe were chiefly to be ascribed to the admirable inventive power and common-sense genius of William Murdock, made me feel that I was indeed on classic ground in regard to everything connected with the construction of steam-engine machinery. The interest was in no small degree enhanced by coming every now and then upon some machine that had every historical claim to be regarded as the prototype of many of our modern machine tools. All these had William Murdock's genius stamped upon them, by reason of their common-sense arrangements, which showed that he was one of those original thinkers who had the courage to break away from the trammels of traditional methods, and take short cuts to accomplish his objects by direct and simple means."

We have another recollection of William Murdock, from one who knew him when a boy. This is the venerable Charles Manby, F.R.S., still honorary secretary of the Institute of Civil Engineers. He says (writing to us in September 1883), "I see from the public prints that you have been presiding at a meeting intended to do honour to the memory of William Murdock—a most worthy man and an old friend of mine. When he found me working the first slide valve ever introduced into an engine-building establishment at Horsley, he patted me on the head, and said to my father, 'Neighbour Manby, this is not the way to bring up a good workman—merely turning a handle, without any shoulder work.' He evidently did not anticipate any great results from my engineering education. But we all know what machine tools are doing now,—and where should we be without them?"

Watt withdrew from the firm in 1800, on the expiry of his patent for the condensing steam-engine; but Boulton continued until the year 1809, when he died full of years and honours. Watt lived on until 1819. The last part of his life was the happiest. During the time that he was in the throes of his invention, he was very miserable, weighed down with dyspepsia and sick headaches. But after his patent had expired, he was able to retire with a moderate fortune, and began to enjoy life. Before, he had "cursed his inventions," now he could bless them. He was able to survey them, and find out what was right and what was wrong. He used his head and his hands in his private workshop, and found many means of employing both pleasantly. Murdock continued to be his fast friend, and they spent many agreeable hours together. They made experiments and devised improvements in machines. Watt wished to make things more simple. He said to Murdock, "it is a great thing to know what to do without. We must have a book of blots—things to be scratched out." One of the most interesting schemes of Watt towards the end of his life was the contrivance of a sculpture-making machine; and he proceeded so far with it as to to able to present copies of busts to his friends as "the productions of a young artist just entering his eighty-third year." The machine, however, remained unfinished at his death, and the remarkable fact is that it was Watt's only unfinished work.

The principle of the machine was to carry a guide-point at one side over the bust or alto-relievo to be copied, and at the other side to carry a corresponding cutting-tool or drill over the alabaster, ivory, jet, or plaster of Paris to be executed. The machine worked, as it were, with two hands, the one feeling the pattern, the other cutting the material into the required form. Many new alterations were necessary for carrying out this ingenious apparatus, and Murdock was always at hand to give his old friend and master his best assistance. We have seen many original letters from Watt to Murdock, asking for counsel and help. In one of these, written in 1808, Watt says: "I have revived an idea which, if it answers, will supersede the frame and upright spindle of the reducing machine, but more of this when we meet. Meanwhile it will be proper to adhere to the frame, etc., at present, until we see how the other alterations answer." In another he says: "I have done a Cicero without any plaits—the different segments meeting exactly. The fitting the drills into the spindle by a taper of 1 in 6 will do. They are perfectly stiff and will not unscrew easily. Four guide-pullies answer, but there must be a pair for the other end, and to work with a single hand, for the returning part is always cut upon some part or other of the frame."

These letters are written sometimes in the morning, sometimes at noon, sometimes at night. There was a great deal of correspondence about "pullies," which did not seem to answer at first. "I have made the tablets," said Watt on one occasion, "slide more easily, and can counterbalance any part of their weight which may be necessary; but the first thing to try is the solidity of the machine, which cannot be done till the pullies are mounted." Then again: "The bust-making must be given up until we get a more solid frame. I have worked two days at one and spoiled it, principally from the want of steadiness." For Watt, it must be remembered, was now a very old man.

He then proceeded to send Murdock the drawing of a "parallel motion for the machine," to be executed by the workmen at Soho. The truss braces and the crosses were to be executed of steel, according to the details he enclosed. "I have warmed up," he concludes, "an old idea, and can make a machine in which the pentagraph and the leading screw will all be contained in the beam, and the pattern and piece to be cut will remain at rest fixed upon a lath of cast iron or stout steel." Watt is very particular in all his details: "I am sorry," he says in one note, "to trouble you with so many things; but the alterations on this spindle and socket [he annexes a drawing] may wait your convenience." In a further note, Watt says. "The drawing for the parallel lathe is ready; but I have been sadly puzzled about the application of the leading screws to the cranes in the other. I think, however, I have now got the better of the difficulties, and made it more certain, as well as more simple, than it was. I have done an excellent head of John Hunter in hard white in shorter time than usual. I want to show it you before I repair it."

At last Watt seems to have become satisfied: "The lathe," he says, "is very much improved, and you seem to have given the finishing blow to the roofed frame, which appears perfectly stiff. I had some hours' intense thinking upon the machine last night, and have made up my mind on it at last. The great difficulty was about the application of the band, but I have settled it to be much as at present."

Watt's letters to Murdock are most particular in details, especially as to screws, nuts, and tubes, with strengths and dimensions, always illustrated with pen-and-ink drawings. And yet all this was done merely for mechanical amusement, and not for any personal pecuniary advantage. While Watt was making experiments as to the proper substances to be carved and drilled, he also desired Murdock to make similar experiments. "The nitre," he said in one note, "seems to do harm; the fluor composition seems the best and hardest. Query, what would some calcined pipe-clay do? If you will calcine some fire-clay by a red heat and pound it,—about a pound,—and send it to me, I shall try to make you a mould or two in Henning's manner to cast this and the sulphur acid iron in. I have made a screwing tool for wood that seems to answer; also one of a one-tenth diameter for marble, which does very well." In another note, Watt says: "I find my drill readily makes 2400 turns per minute, even with the large drill you sent last; if I bear lightly, a three-quarter ferril would run about 3000, and by an engine that might be doubled."

The materials to be drilled into medallions also required much consideration. "I am much obliged to you," said Watt, "for the balls, etc., which answer as well as can be expected. They make great progress in cutting the crust (Ridgways) or alabaster, and also cut marble, but the harder sorts soon blunt them. At any rate, marble does not do for the medallions, as its grain prevents its being cut smooth, and its semi-transparence hurts the effect. I think Bristol lime, or shell lime, pressed in your manner, would have a good effect. When you are at leisure, I shall thank you for a few pieces, and if some of them are made pink or flesh colour, they will look well. I used the ball quite perpendicular, and it cut well, as most of the cutting is sideways. I tried a fine whirling point, but it made little progress; another with a chisel edge did almost as well as the balls, but did not work so pleasantly. I find a triangular scraping point the best, and I think from some trials it should be quite a sharp point. The wheel runs easier than it did, but has still too much friction. I wished to have had an hour's consultation with you, but have been prevented by sundry matters among others by that plaguey stove, which is now in your hands."

Watt was most grateful to Murdock for his unvarying assistance. In January, 1813, when Watt was in his seventy-seventh year, he wrote to Murdock, asking him to accept a present of a lathe "I have not heard from you," he says, "in reply to my letter about the lathe; and, presuming you are not otherwise provided, I have bought it, and request your acceptance of it. At present, an alteration for the better is making in the oval chuck, and a few additional chucks, rest, etc., are making to the lathe. When these are finished, I shall have it at Billinger's until you return, or as you otherwise direct. I am going on with my drawings for a complete machine, and shall be glad to see you here to judge of them."

The drawings were made, but the machine was never finished. "Invention," said Watt, "goes on very slowly with me now." Four years later, he was still at work; but death put a stop to his "diminishing-machine." It is a remarkable testimony to the skill and perseverance of a man who had already accomplished so much, that it is almost his only unfinished work. Watt died in 1819, in the eighty-third year of his age, to the great grief of Murdock, his oldest and most attached friend and correspondent.

Meanwhile, the firm of Boulton and Watt continued. The sons of the two partners carried it on, with Murdock as their Mentor. He was still full of work and inventive power. In 1802, he applied the compressed air of the Blast Engine employed to blow the cupolas of the Soho Foundry, for the purpose of driving the lathe in the pattern shop. It worked a small engine, with a 12-inch cylinder and 18-inch stroke, connected with the lathe, the speed being regulated as required by varying the admission of the blast. This engine continued in use for about thirty-five years.

In 1803 Murdock experimented on the power of high-pressure steam in propelling shot, and contrived a steam-engine with which he made many trials at Soho, thereby anticipating the apparatus contrived by Mr. Perkins many years later.

In 1810 Murdock took out a patent for boring steam-pipes for water, and cutting columns out of solid blocks of stone, by means of a cylindrical crown saw. The first machine was used at Soho, and afterwards at Mr. Rennie's Works in London, and proved quite successful. Among his other inventions were a lift worked by compressed air, which raised and lowered the castings from the boring-mill to the level of the foundry and the canal bank. He used the same kind of power to ring the bells in his house at Sycamore Hill, and the contrivance was afterwards adopted by Sir Walter Scott in his house at Abbotsford.

Murdock was also the inventor of the well-known cast-iron cement, so extensively used in engine and machine work. The manner in which he was led to this invention affords a striking illustration of his quickness of observation. Finding that some iron-borings and sal-ammoniac had got accidently mixed together in his tool-chest, and rusted his saw-blade nearly through, he took note of the circumstance, mixed the articles in various proportions, and at length arrived at the famous cement, which eventually became an article of extensive manufacture at the Soho Works.

Murdock's ingenuity was constantly at work, even upon matters which lay entirely outside his special vocation. The late Sir William Fairbairn informed us that he contrived a variety of curious machines for consolidating peat moss, finely ground and pulverised, under immense pressure, and which, when consolidated, could be moulded into beautiful medals, armlets, and necklaces. The material took the most brilliant polish and had the appearance of the finest jet.

Observing that fish-skins might be used as an economical substitute for isinglass, he went up to London on one occasion in order to explain to brewers the best method of preparing and using them. He occupied handsome apartments, and, little regarding the splendour of the drawing-room, he hung the fish-skins up against the walls. His landlady caught him one day when he was about to bang up a wet cod's skin! He was turned out at once, with all his fish. While in town on this errand, it occurred to him that a great deal of power was wasted in treading the streets of London! He conceived the idea of using the streets and roadways as a grand tread-mill, under which the waste power might be stored up by mechanical methods and turned to account. He had also an idea of storing up the power of the tides, and of running water, in the same way. The late

Charles Babbage, F.R.S., entertained a similar idea about using springs of Ischia or of the geysers of Iceland as a power necessary for condensing gases, or perhaps for the storage of electricity.[12] The latter, when perfected, will probably be the greatest invention of the next half century.

Another of Murdock's' ingenious schemes, was his proposed method of transmitting letters and packages through a tube exhausted by an air-pump. This project led to the Atmospheric Railway, the success of which, so far as it went, was due to the practical ability of Murdock's pupil, Samuel Clegg. Although the atmospheric railway was eventually abandoned, it is remarkable that the original idea was afterwards revived and practised with success by the London Pneumatic Dispatch Company.

In 1815, while Murdock was engaged in erecting an apparatus of his own invention for heating the water for the baths at Leamington, a ponderous cast-iron plate fell upon his leg above his ankle, and severely injured him. He remained a long while at Leamington, and when it was thought safe to remove him, the Birmingham Canal Company kindly placed their excursion boat at his disposal, and he was conveyed safely homeward. So soon as he was able, he was at work again at the Soho factory.

Although the elder Watt had to a certain extent ignored the uses of steam as applied to navigation, being too much occupied with developing the powers of the pumping and rotary engine, the young partners, with the stout aid of Murdock, took up the question. They supplied Fulton in 1807 with his first engine, by means of which the Clermont made her first voyage along the Hudson river. They also supplied Fulton and Livingston with the next two engines for the Car of Neptune and the Paragon. From that time forward, Boulton and Watt devoted themselves to the manufacture of engines for steamboats. Up to the year 1814, marine engines had been all applied singly in the vessel; but in this year Boulton and Watt first applied two condensing engines, connected by cranks set at right angles on the shaft, to propel a steamer on the Clyde. Since then, nearly all steamers are fitted with two engines. In making this important improvement, the firm were materially aided by the mechanical genius of William Murdock, and also of Mr. Brown, then an assistant, but afterwards a member of the firm.

In order to carry on a set of experiments with respect to the most improved form of marine engine, Boulton and Watt purchased the Caledonia, a Scotch boat built on the Clyde by James Wood and Co., of Port Glasgow. The engines and boilers were taken out. The vessel was fitted with two side lever engines, and many successive experiments were made with her down to August, 1817, at an expense of about 10,000L. This led to a settled plan of construction, by which marine engines were greatly improved. James Watt, junior, accompanied the Caledonia to Holland and up the Rhine. The vessel was eventually sold to the Danish Government, and used for carrying the mails between Kiel and Copenhagen. It is, however, unnecessary here to venture upon the further history of steam navigation.

In the midst of these repeated inventions and experiments, Murdock was becoming an old man. Yet he never ceased to take an interest in the works at Soho. At length his faculties experienced a gradual decay, and he died peacefully at his house at Sycamore Hill, on the 15th of November,1839, in his eighty-fifth year. He was buried near the remains of the great Boulton and Watt; and a bust by Chantrey served to perpetuate the remembrance of his manly and intelligent countenance.

Footnotes for Chapter V.

[1] Fletcher's Political Works, London, 1737, p. 149,

[2] One of the Murdocks built the cathedral at Glasgow, as well as others in Scotland. The famous school of masonry at Antwerp sent out a number of excellent architects during the 11th, 12th, and 13th centuries. One of these, on coming into Scotland, assumed the name of Murdo. He was a Frenchman, born in Paris, as we learn from the inscription left on Melrose Abbey, and he died while building that noble work: it is as follows:—

"John Murdo sumtyme cait was I And born in Peryse certainly, An' had in kepyng all mason wark Sanct Andrays, the Hye Kirk o' Glasgo, Melrose and Paisley, Jedybro and Galowy. Pray to God and Mary baith, and sweet Saint John, keep this Holy Kirk frae scaith."

[3] The discovery of the Black Band Ironstone by David Mushet in 1801, and the invention of the Hot Blast by James Beaumont Neilson in 1828, will be found related in Industrial Biography, pp. 141-161.

[4] Note to Lockhart's Life of Scott.

[5] This was stated to the present writer some years ago by William Murdock's son; although there is no other record of the event.

[6] See Lives of Engineers (Boulton and Watt), iv. pp. 182-4. Small edition, pp. 130-2.

[7] Mr. Pearse's letter is dated 23rd April, 1867, but has not before been published. He adds that "others remembered Murdock, one who was an apprentice with him, and lived with him for some time—a Mr. Vivian, of the foundry at Luckingmill."

[8] Murdock's house still stands in Cross Street, Redruth; those still live who saw the gas-pipes conveying gas from the retort in the little yard to near the ceiling of the room, just over the table; a hole for the pipe was made in the window frame. The old window is now replaced by a new frame."—Life of Richard Trevithick, i. 64.

[9] Philosophical Transactions, 1808, pp. 124-132.

[10] Winsor's family evidently believed in his great powers; for I am informed by Francis Galton, Esq., F.R.S., that there is a fantastical monument on the right-hand side of the central avenue of the Kensal Green Cemetery, about

half way between the lodge and the church, which bears the following inscription:—"Tomb of Frederick Albert Winsor, son of the late Frederick Albert Winsor, originator of public Gas-lighting, buried in the Cemetery of Pere la Chaise, Paris. At evening time it shall be light."—Zachariah xiv. 7. "I am come a light into the world, that whoever believeth in Me shall not abide in darkness."—John xii. 46.

[11] Mr. Parkes, in his well known Chemical Essays (ed. 1841, p. 157), after referring to the successful lighting up by Murdock of the manufactory of Messrs. Phillips and Lee at Manchester in 1805, "with coal gas issuing from nearly a thousand burners," proceeds, "This grand application of the new principle satisfied the public mind, not only of the practicability, but also of the economy of the application; and as a mark of the high opinion they entertained of his genius and perseverance, and in order to put the question of priority of the discovery beyond all doubt, the Council of the Royal Society in 1808 awarded to Mr. Murdock the Gold Medal founded by the late Count Rumford."

[12] "Thus," says Mr. Charles Babbage, "in a future age, power may become the staple commodity of the Icelanders, and of the inhabitants of other volcanic districts; and possibly the very process by which they will procure this article of exchange for the luxuries of happier climates may, in some measure, tame the tremendous element which occasionally devastates their provinces."—Economy of Manufactures.

CHAPTER VI.

FREDERICK KOENIG: INVENTOR OF THE STEAM-PRINTING MACHINE.

"The honest projector is he who, having by fair and plain principles of sense, honesty, and ingenuity, brought any contrivance to a suitable perfection, makes out what he pretends to, picks nobody's pocket, puts his project in execution, and contents himself with the real produce as the profit of his invention."—De Foe.

I published an article in 'Macmillan's Magazine' for December, 1869, under the above title. The materials were principally obtained from William and Frederick Koenig, sons of the inventor.

Since then an elaborate life has been published at Stuttgart, under the title of "Friederich Koenig und die Erfindung Der Schnellpresse, Ein Biographisches Denkmal. Von Theodor Goebel." The author, in sending me a copy of the volume, refers to the article published in 'Macmillan,' and says, "I hope you will please to accept it as a small acknowledgment of the thanks, which every German, and especially the sons of Koenig, in whose name I send the book as well as in mine, owe to you for having bravely taken up the cause of the much wronged inventor, their father—an action all the more praiseworthy, as you had to write against the prejudices and the interests of your own countrymen."

I believe it is now generally admitted that Koenig was entitled to the merit of being the first person practically to apply the power of steam to indefinitely multiplying the productions of the printing-press; and that no one now attempts to deny him this honour. It is true others, who followed him, greatly improved upon his first idea; but this was the case with Watt, Symington, Crompton, Maudslay, and many more. The true inventor is not merely the man who registers an idea and takes a patent for it, or who compiles an invention by borrowing the idea of another, improving upon or adding to his arrangements, but the man who constructs a machine such as has never before been made, which executes satisfactorily all the functions it was intended to perform. And this is what Koenig's invention did, as will be observed from the following brief summary of his life and labours.

Frederick Koenig was born on the 17th of April, 1774, at Eisleben, in Saxony, the birthplace also of a still more famous person, Martin Luther. His father was a respectable peasant proprietor, described by Herr Goebel as Anspanner. But this word has now gone out of use. In feudal times it described the farmer who was obliged to keep draught cattle to perform service due to the landlord. The boy received a solid education at the Gymnasium, or public school of the town. At a proper age he was bound apprentice for five years to Breitkopf and Hartel, of Leipzig, as compositor and printer; but after serving for four and a quarter years, he was released from his engagement because of his exceptional skill, which was an unusual occurrence.

During the later years of his apprenticeship, Koenig was permitted to attend the classes in the University, more especially those of Ernst Platner, a physician, philosopher, and anthropologist. After that he proceeded to the printing-office of his uncle, Anton F. Rose, at Greifswald, an old seaport town on the Baltic, where he remained a few years. He next went to Halle as a journeyman printer,—German workmen going about from place to place, during their wanderschaft, for the purpose of learning their business. After that, he returned to Breitkopf and Hartel, at Leipzig, where he had first learnt his trade. During this time, having saved a little money, he enrolled himself for a year as a regular student at the University of Leipzig.

According to Koenig's own account, he first began to devise ways and means for improving the art of printing in the year 1802, when he was twenty-eight years old. Printing large sheets of paper by hand was a very slow as well as a very laborious process. One of the things that most occupied the young printer's mind was how to get rid of this "horse-work," for such it was, in the business of printing. He was not, however, over-burdened with means, though he devised a machine with this object. But to make a little money, he made translations for the publishers. In 1803 Koenig returned to his native town of Eisleben, where he entered into an arrangement with Frederick Riedel, who furnished the necessary capital for carrying on the business of a printer and bookseller. Koenig alleges that his reason for adopting this step was to raise sufficient money to enable him to carry out his plans for the improvement of printing.

The business, however, did not succeed, as we find him in the following year carrying on a printing trade at Mayence. Having sold this business, he removed to Suhl in Thuringia. Here he was occupied with a stereotyping process, suggested by what he had read about the art as perfected in England by Earl Stanhope. He also contrived an improved press, provided with a moveable carriage, on which the types were placed, with inking rollers, and a new mechanical method of taking off the impression by flat pressure.

Koenig brought his new machine under the notice of the leading printers in Germany, but they would not undertake to use it. The plan seemed to them too complicated and costly. He tried to enlist men of capital in his scheme, but they all turned a deaf ear to him. He went from town to town, but could obtain no encouragement whatever. Besides, industrial enterprise in Germany was then in a measure paralysed by the impending war with France, and men of capital were naturally averse to risk their money on what seemed a merely speculative undertaking.

Finding no sympathisers or helpers at home, Koenig next turned his attention abroad. England was then, as now, the refuge of inventors who could not find the means of bringing out their schemes elsewhere; and to England he wistfully turned his eyes. In the meantime, however, his inventive ability having become known, an offer was made to him by the Russian Government to proceed to St. Petersburg and organise the State printing-office there. The invitation was accepted, and Koenig proceeded to St. Petersburg in the spring of 1806. But the official difficulties thrown in his way were very great, and so disgusted him, that he decided to throw up his appointment, and try his fortune in England. He accordingly took ship for London, and arrived there in the following November, poor in means, but rich in his great idea, then his only property.

As Koenig himself said, when giving an account of his invention:—"There is on the Continent no sort of encouragement for an enterprise of this description. The system of patents, as it exists in England, being either unknown, or not adopted in the Continental States, there is no inducement for industrial enterprise; and projectors are commonly obliged to offer their discoveries to some Government, and to so licit their encouragement. I need hardly add that scarcely ever is an invention brought to maturity under such circumstances. The well-known fact, that almost every invention seeks, as it were, refuge in England, and is there brought to perfection, though the Government does not afford any other protection to inventors beyond what is derived from the wisdom of the laws, seems to indicate that the Continent has yet to learn from her the best manner of encouraging the mechanical arts. I had my full share in the ordinary disappointments of Continental projectors; and after having lost in Germany and Russia upwards of two years in fruitless applications, I at last resorted to England."[1]

After arriving in London, Koenig maintained himself with difficulty by working at his trade, for his comparative ignorance of the English language stood in his way. But to work manually at the printer's "case," was not Koenig's object in coming to England. His idea of a printing machine was always uppermost in his mind, and he lost no opportunity of bringing the subject under the notice of master printers likely to take it up. He worked for a time in the printing office of Richard Taylor, Shoe Lane, Fleet Street, and mentioned the matter to him. Taylor would not undertake the invention himself, but he furnished Koenig with an introduction to Thomas Bensley, the well-known printer of Bolt Court, Fleet Street. On the 11th of March, 1807, Bensley invited Koenig to meet him on the subject of their recent conversation about "the discovery;" and on the 31st of the same month, the following agreement was entered into between Koenig and Bensley:—

"Mr. Koenig, having discovered an entire new Method of Printing by Machinery, agrees to communicate the same to Mr. Bensley under the following conditions:—that, if Mr. Bensley shall be satisfied the Invention will answer all the purposes Mr. Koenig has stated in the Particulars he has delivered to Mr. Bensley, signed with his name, he shall enter into a legal Engagement to purchase the Secret from Mr. Koenig, or enter into such other agreement as may be deemed mutually beneficial to both parties; or, should Mr. Bensley wish to decline having any concern with the said Invention, then he engages not to make any use of the Machinery, or to communicate the Secret to any person whatsoever, until it is proved that the Invention is made use of by any one without restriction of Patent, or other particular agreement on the part of Mr. Koenig, under the penalty of Six Thousand Pounds.

"(Signed) T. Bensley,
"Friederich Konig.
"Witness—J. Hunneman."

Koenig now proceeded to put his idea in execution. He prepared his plans of the new printing machine. It seems, however, that the progress made by him was very slow. Indeed, three years passed before a working model could be got ready, to show his idea in actual practice. In the meantime, Mr. Walter of The Times had been seen by Bensley, and consulted on the subject of the invention. On the 9th of August, 1809, more than two years after the date of the above agreement, Bensley writes to Koenig: "I made a point of calling upon Mr. Walter yesterday, who, I am sorry to say, declines our proposition altogether, having (as he says) so many engagements as to prevent him entering into more."

It may be mentioned that Koenig's original plan was confined to an improved press, in which the operation of laying the ink on the types was to be performed by an apparatus connected with the motions of the coffin, in such a manner as that one hand could be saved. As little could be gained in expedition by this plan, the idea soon suggested itself of moving the press by machinery, or to reduce the several operations to one rotary motion, to which the first mover might be applied. Whilst Koenig was in the throes of his invention, he was joined by his friend

Andrew F. Bauer, a native of Stuttgart, who possessed considerable mechanical power, in which the inventor himself was probably somewhat deficient. At all events, these two together proceeded to work out the idea, and to construct the first actual working printing machine.

A patent was taken out, dated the 29th of March, 1810, which describes the details of the invention. The arrangement was somewhat similar to that known as the platen machine; the printing being produced by two flat plates, as in the common hand-press. It also embodied an ingenious arrangement for inking the type. Instead of the old-fashioned inking balls, which were beaten on the type by hand labour, several cylinders covered with felt and leather were used, and formed part of the machine itself. Two of the cylinders revolved in opposite directions, so as to spread the ink, which was then transferred by two other inking cylinders alternately applied to the "forme" by the action of spiral springs. The movement of all the parts of the machine were to be derived from a steam-engine, or other first mover.

"After many obstructions and delays," says Koenig himself, in describing the history of his invention, "the first printing machine was completed exactly upon the plan which I have described in the specification of my first patent. It was set to Work in April, 1811. The sheet (H) of the new Annual Register for 1810, 'Principal Occurrences,' 3000 copies, was printed with it; and is, I have no doubt, the first part of a book ever printed with a machine. The actual use of it, however, soon suggested new ideas, and led to the rendering it less complicated and more powerful"[2]

Of course! No great invention was ever completed at one effort. It would have been strange if Koenig had been satisfied with his first attempt. It was only a beginning, and he naturally proceeded with the improvement of his machine. It took Watt more than twenty years to elaborate his condensing steam-engine; and since his day, owing to the perfection of self-acting tools, it has been greatly improved. The power of the Steamboat and the Locomotive also, as well as of all other inventions, have been developed by the constantly succeeding improvements of a nation of mechanical engineers.

Koenig's experiment was only a beginning, and he naturally proceeded with the improvement of his machine. Although the platen machine of Koenig's has since been taken up a new, and perfected, it was not considered by him sufficiently simple in its arrangements as to be adapted for common use; and he had scarcely completed it, when he was already revolving in his mind a plan of a second machine on a new principle, with the object of ensuring greater speed, economy, and simplicity.

By this time, other well-known London printers, Messrs. Taylor and Woodfall, had joined Koenig and Bensley in their partnership for the manufacture and sale of printing machines. The idea which now occurred to Koenig was, to employ a cylinder instead of a flat Platen machine, for taking the impressions off the type, and to place the sheet round the cylinder, thereby making it, as it were, part of the periphery. As early as the year 1790, one William Nicholson had taken out a patent for a machine for printing "on paper, linen, cotton, woollen, and other articles," by means of "blocks, forms, types, plates, and originals," which were to be "firmly imposed upon a cylindrical surface in the same manner as common letter is imposed upon a flat stone."[3] From the mention of "colouring cylinder," and "paper-hangings, floor-cloths, cottons, linens, woollens, leather, skin, and every other flexible material," mentioned in the specification, it would appear as if Nicholson's invention were adapted for calico-printing and paper-hangings, as well as for the printing of books. But it was never used for any of these purposes. It contained merely the register of an idea, and that was all. It was left for Adam Parkinson, of Manchester, to invent and make practical use of the cylinder printing machine for calico in the year 1805, and this was still further advanced by the invention of James Thompson, of Clitheroe, in 1813; while it was left for Frederick Koenig to invent and carry into practical operation the cylinder printing press for newspapers.

After some promising experiments, the plans for a new machine on the cylindrical principle were proceeded with. Koenig admitted throughout the great benefit he derived from the assistance of his friend Bauer. "By the judgment and precision," he said, "with which he executed my plans, he greatly contributed to my success." A patent was taken out on October 30th, 1811; and the new machine was completed in December, 1812. The first sheets ever printed with an entirely cylindrical press, were sheets G and X of Clarkson's 'Life of Penn.' The papers of the Protestant Union were also printed with it in February and March, 1813. Mr. Koenig, in his account of the invention, says that "sheet M of Acton's 'Hortus Kewensis,' vol. v., will show the progress of improvement in the use of the invention. Altogether, there are about 160,000 sheets now in the hands of the public, printed with this machine, which, with the aid of two hands, takes off 800 impressions in the hour"[4]

Koenig took out a further patent on July 23rd, 1813, and a fourth (the last) on the 14th of March, 1814. The contrivance of these various arrangements cost the inventor many anxious days and nights of study and labour. But he saw before him only the end he wished to compass, and thought but little of himself and his toils. It may be mentioned that the principal feature of the invention was the printing cylinder in the centre of the machine, by which the impression was taken from the types, instead of by flat plates as in the first arrangement. The forme was fixed in a cast-iron plate which was carried to and fro on a table, being received at either end by strong spiral springs. A double machine, on the same principle,—the forme alternately passing under and giving an impression at one of two cylinders at either end of the press,—was also included in the patent of 1811.

How diligently Koenig continued to elaborate the details of his invention will be obvious from the two last patents which he took out, in 1813 and 1814. In the first he introduced an important improvement in the inking arrangement, and a contrivance for holding and carrying on the sheet, keeping it close to the printing cylinder by means of endless tapes; while in the second, he added the following new expedients: a feeder, consisting of an endless web,—an improved arrangement of the endless tapes by inner as well as outer friskets,—an improvement of the register (that is, one page falling exactly on the back of another), by which greater accuracy of impression was also secured; and finally, an arrangement by which the sheet was thrown out of the machine, printed by the revolving cylinder on both sides.

The partners in Koenig's Patents had established a manufactory in Whitecross Street for the production of the new machines. The workmen employed were sworn to secrecy. They entered into an agreement by which they were liable to forfeit 100L. if they communicated to others the secret of the machines, either by drawings or description, or if they told by whom or for whom they were constructed. This was to avoid the hostility of the pressmen, who, having heard of the new invention, were up in arms against it, as likely to deprive them of their employment. And yet, as stated by Johnson in his 'Typographia,' the manual labour of the men who worked at the hand press, was so severe and exhausting, "that the stoutest constitutions fell a sacrifice to it in a few years." The number of sheets that could be thrown off was also extremely limited.

With the improved press, perfected by Earl Stanhope, about 250 impressions could be taken, or 125 sheets printed on both sides in an hour. Although a greater number was produced in newspaper printing offices by excessive labour, yet it was necessary to have duplicate presses, and to set up duplicate forms of type, to carry on such extra work; and still the production of copies was quite inadequate to satisfy the rapidly increasing demand for newspapers. The time was therefore evidently ripe for the adoption of such a machine as that of Koenig. Attempts had been made by many inventors, but every one of them had failed. Printers generally regarded the steam-press as altogether chimerical.

Such was the condition of affairs when Koenig finished his improved printing machine in the manufactory in Whitecross Street. The partners in the invention were now in great hopes. When the machine had been got ready for work, the proprietors of several of the leading London newspapers were invited to witness its performances. Amongst them were Mr. Perry of the Morning chronicle, and Mr. Walter of The Times. Mr. Perry would have nothing to do with the machine; he would not even go to see it, for he regarded it as a gimcrack.[5] On the contrary, Mr. Walter, though he had five years before declined to enter into any arrangement with Bensley, now that he heard the machine was finished, and at work, decided to go and inspect it. It was thoroughly characteristic of the business spirit of the man. He had been very anxious to apply increased mechanical power to the printing of his newspaper. He had consulted Isambard Brunel—one of the cleverest inventors of the day—on the subject; but Brunel, after studying the subject, and labouring over a variety of plans, finally gave it up. He had next tried Thomas Martyn, an ingenious young compositor, who had a scheme for a self-acting machine for working the printing press. But, although Mr. Walter supplied him with the necessary funds, his scheme never came to anything. Now, therefore, was the chance for Koenig!

After carefully examining the machine at work, Mr. Walter was at once satisfied as to the great value of the invention. He saw it turning out the impressions with unusual speed and great regularity. This was the very machine of which he had been in search. But it turned out the impressions printed on one side only. Koenig, however, having briefly explained the more rapid action of a double machine on the same principle for the printing of newspapers, Mr. Walter, after a few minutes' consideration, and before leaving the premises, ordered two double machines for the printing of The Times newspaper. Here, at last, was the opportunity for a triumphant issue out of Koenig's difficulties.

The construction of the first newspaper machine was still, however, a work of great difficulty and labour. It must be remembered that nothing of the kind had yet been made by any other inventor. The single-cylinder machine, which Mr. Walter had seen at work, was intended for bookwork only. Now Koenig had to construct a double-cylinder machine for printing newspapers, in which many of the arrangements must necessarily be entirely new. With the assistance of his leading mechanic, Bauer, aided by the valuable suggestions of Mr. Walter himself, Koenig at length completed his plans, and proceeded with the erection of the working machine. The several parts were prepared at the workshop in Whitecross Street, and taken from thence, in as secret a way as possible, to the premises in Printing House Square, adjoining The Times office, where they were fitted together and erected into a working machine. Nearly two years elapsed before the press was ready for work. Great as was the secrecy with which the operations were conducted, the pressmen of The Times office obtained some inkling of what was going on, and they vowed vengeance to the foreign inventor who threatened their craft with destruction. There was, however, always this consolation: every attempt that had heretofore been made to print newspapers in any other way than by manual labour had proved an utter failure!

At length the day arrived when the first newspaper steam-press was ready for use. The pressmen were in a state of great excitement, for they knew by rumour that the machine of which they had so long been apprehensive was fast approaching completion. One night they were told to wait in the press-room, as important news was expected from abroad. At six o'clock in the morning of the 29th November, 1814, Mr. Walter, who had been watching the working of the machine all through the night, suddenly appeared among the pressmen, and announced that "The Times is already printed by steam!" Knowing that the pressmen had vowed vengeance against the inventor and his invention, and that they had threatened "destruction to him and his traps," he informed them that if they attempted violence, there was a force ready to suppress it; but that if they were peaceable, their wages should be continued to every one of them until they could obtain similar employment. This proved satisfactory so far, and he proceeded to distribute several copies of the newspaper amongst them—the first newspaper printed by steam! That paper contained the following memorable announcement:—

"Our Journal of this day presents to the Public the practical result of the greatest improvement connected with printing since the discovery of the art itself. The reader of this paragraph now holds in his hand one of the many thousand impressions of The Times newspaper which were taken off last night by a mechanical apparatus. A system of machinery almost organic has been devised and arranged, which, while it relieves the human frame of its most laborious' efforts in printing, far exceeds all human powers in rapidity and dispatch. That the magnitude of the invention may be justly appreciated by its effects, we shall inform the public, that after the letters are placed by the compositors, and enclosed in what is called the forme, little more remains for man to do than to attend upon and to watch this unconscious agent in its operations. The machine is then merely supplied with paper: itself places the forme, inks it, adjusts the paper to the forme newly inked, stamps the sheet, and gives it forth to the hands of the attendant, at the same time withdrawing the forme for a fresh coat of ink, which itself again distributes, to meet the ensuing sheet now advancing for impression; and the whole of these complicated acts is performed with such a velocity and simultaneousness of movement, that no less than 1100 sheets are impressed in one hour.

"That the completion of an invention of this kind, not the effect of chance, but the result of mechanical combinations methodically arranged in the mind of the artist, should be attended with many obstructions and much delay, may be readily imagined. Our share in this event has, indeed, only been the application of the discovery, under an agreement with the patentees, to our own particular business; yet few can conceive—even with this limited interest—the various disappointments and deep anxiety to which we have for a long course of time been subjected.

"Of the person who made this discovery we have but little to add. Sir Christopher Wren's noblest monument is to be found in the building which he erected; so is the best tribute of praise which we are capable of offering to the inventor of the printing machine, comprised in the preceding description, which we have feebly sketched, of the powers and utility of his invention. It must suffice to say further, that he is a Saxon by birth; that his name is Koenig; and that the invention has been executed under the direction of his friend and countryman, Bauer."

The machine continued to work steadily and satisfactorily, notwithstanding the doubters, the unbelievers, and the threateners of vengeance. The leading article of The Times for December 3rd, 1814, contains the following statement:—

"The machine of which we announced the discovery and our adoption a few days ago, has been whirling on its course ever since, with improving order, regularity, and even speed. The length of the debates on Thursday, the day when Parliament was adjourned, will have been observed; on such an occasion the operation of composing and printing the last page must commence among all the journals at the same moment; and starting from that moment, we, with our infinitely superior circulation, were enabled to throw off our whole impression many hours before the other respectable rival prints. The accuracy and clearness of the impression will likewise excite attention.

"We shall make no reflections upon those by whom this wonderful discovery has been opposed,—the doubters and unbelievers,—however uncharitable they may have been to us; were it not that the efforts of genius are always impeded by drivellers of this description, and that we owe it to such men as Mr. Koenig and his Friend, and all future promulgators of beneficial inventions, to warn them that they will have to contend with everything that selfishness and conceited ignorance can devise or say; and if we cannot clear their way before them, we would at least give them notice to prepare a panoply against its dirt and filth.

"There is another class of men from whom we receive dark and anonymous threats of vengeance if we persevere in the use of this machine. These are the Pressmen. They well know, at least should well know, that such menace is thrown away upon us. There is nothing that we will not do to assist and serve those whom we have discharged. They themselves can seethe greater rapidity and precision with which the paper is printed. What right have they to make us print it slower and worse for their supposed benefit? A little reflection, indeed, would show them that it is neither in their power nor in ours to stop a discovery now made, if it is beneficial to mankind; or to force it down if it is useless. They had better, therefore, acquiesce in a result which they cannot alter; more especially as there will still be employment enough for the old race of pressmen, before the new method obtains general use, and no new ones need be brought up to the business; but we caution them seriously against involving themselves and their families in ruin, by becoming amenable to the laws of their country. It has always been matter of great satisfaction to us to reflect, that we encountered and crushed one conspiracy; and we should be sorry to find our work half done.

"It is proper to undeceive the world in one particular; that is, as to the number of men discharged. We in fact employ only eight fewer workmen than formerly; whereas more than three times that number have been employed for a year and a half in building the machine."

On the 8th of December following, Mr. Koenig addressed an advertisement "To the Public" in the columns of The Times, giving an account of the origin and progress of his invention. We have already cited several passages from the statement. After referring to his two last patents, he says: "The machines now printing The Times and Mail are upon the same principle; but they have been contrived for the particular purpose of a newspaper of extensive circulation, where expedition is the great object.

"The public are undoubtedly aware, that never, perhaps, was a new invention put to so severe a trial as the present one, by being used on its first public introduction for the printing of newspapers, and will, I trust, be indulgent with respect to the many defects in the performance, though none of them are inherent in the principle of the machine; and we hope, that in less than two months, the whole will be corrected by greater adroitness in the management of it, so far at least as the hurry of newspaper printing will at all admit.

"It will appear from the foregoing narrative, that it was incorrectly stated in several newspapers, that I had sold my interest to two other foreigners; my partners in this enterprise being at present two Englishmen, Mr. Bensley and Mr. Taylor; and it is gratifying to my feelings to avail myself of this opportunity to thank those gentlemen publicly for the confidence which they have reposed in me, for the aid of their practical skill, and for the persevering support which they have afforded me in long and very expensive experiments; thus risking their fortunes in the prosecution of my invention.

"The first introduction of the invention was considered by some as a difficult and even hazardous step. The Proprietor of The Times having made that his task, the public are aware that it is in good hands."

One would think that Koenig would now feel himself in smooth water, and receive a share of the good fortune which he had so laboriously prepared for others. Nothing of the kind! His merits were disputed; his rights were denied; his patents were infringed; and he never received any solid advantages for his invention, until he left the country and took refuge in Germany. It is true, he remained for a few years longer, in charge of the manufactory in Whitecross Street, but they were years to him of trouble and sorrow.

In 1816, Koenig designed and superintended the construction of a single cylinder registering machine for book-printing. This was supplied to Bensley and Son, and turned out 1000 sheets, printed on both sides, in the hour. Blumenbach's 'Physiology' was the first entire book printed by steam, by this new machine. It was afterwards employed, in 1818, in working off the Literary Gazette. A machine of the same kind was supplied to Mr. Richard Taylor for the purpose of printing the 'Philosophical Magazine,' and books generally. This was afterwards altered to a double machine, and employed for printing the Weekly Dispatch.

But what about Koenig's patents? They proved of little use to him. They only proclaimed his methods, and enabled other ingenious mechanics to borrow his adaptations. Now that he had succeeded in making machines that would work, the way was clear for everybody else to follow his footsteps. It had taken him more than six years to invent and construct a successful steam printing press; but any clever mechanic, by merely studying his specification, and examining his machine at work, might arrive at the same results in less than a week.

The patents did not protect him. New specifications, embodying some modification or alteration in detail, were lodged by other inventors and new patents taken out. New printing machines were constructed in defiance of his supposed legal rights; and he found himself stripped of the reward that he had been labouring for during so many long and toilsome years. He could not go to law, and increase his own vexation and loss. He might get into Chancery easy enough; but when would he get out of it, and in what condition?

It must also be added, that Koenig was unfortunate in his partner Bensley. While the inventor was taking steps to push the sale of his book-printing machines among the London printers, Bensley, who was himself a book-printer, was hindering him in every way in his negotiations. Koenig was of opinion that Bensley wished to retain the exclusive advantage which the possession of his registering book machine gave him over the other printers, by enabling him to print more quickly and correctly than they could, and thus give him an advantage over them in his printing contracts.

When Koenig, in despair at his position, consulted counsel as to the infringement of his patent, he was told that he might institute proceedings with the best prospect of success; but to this end a perfect agreement by the partners was essential. When, however, Koenig asked Bensley to concur with him in taking proceedings in defence of the patent right, the latter positively refused to do so. Indeed, Koenig was under the impression that his partner had even entered into an arrangement with the infringers of the patent to share with them the proceeds of their piracy.

Under these circumstances, it appeared to Koenig that only two alternatives remained for him to adopt. One was to commence an expensive, and it might be a protracted, suit in Chancery, in defence of his patent rights, with possibly his partner, Bensley, against him; and the other, to abandon his invention in England without further struggle, and settle abroad. He chose the latter alternative, and left England finally in August, 1817.

Mr. Richard Taylor, the other partner in the patent, was an honourable man; but he could not control the proceedings of Bensley. In a memoir published by him in the 'Philosophical Magazine,' "On the Invention and First Introduction of Mr. Koenig's Printing Machine," in which he honestly attributes to him the sole merit of the invention, he says, "Mr. Koenig left England, suddenly, in disgust at the treacherous conduct of Bensley, always shabby and overreaching, and whom he found to be laying a scheme for defrauding his partners in the patents of all the advantages to arise from them. Bensley, however, while he destroyed the prospects of his partners, outwitted himself, and grasping at all, lost all, becoming bankrupt in fortune as well as in character."[6]

Koenig was badly used throughout. His merits as an inventor were denied. On the 3rd of January, 1818, after he had left England, Bensley published a letter in the Literary Gazette, in which he speaks of the printing machine as his own, without mentioning a word of Koenig. The 'British Encyclopaedia,' in describing the inventors of the printing machine, omitted the name of Koenig altogether. The 'Mechanics Magazine,' for September, 1847, attributed the invention to the Proprietors of The Times, though Mr. Walter himself had said that his share in the event had been "only the application of the discovery;" and the late Mr. Bennet Woodcroft, usually a fair man, in his introductory chapter to 'Patents for Inventions in Printing,' attributes the merit to William Nicholson's patent (No. 1748), which, he said, "produced an entire revolution in the mechanism of the art." In other publications, the claims of Bacon and Donkin were put forward, while those of the real inventor were ignored. The memoir of Koenig by Mr. Richard Taylor, in the 'Philosophical Magazine,' was honest and satisfactory; and should have set the question at rest.

It may further be mentioned that William Nicholson,—who was a patent agent, and a great taker out of patents, both in his own name and in the names of others,—was the person employed by Koenig as his agent to take the requisite steps for registering his invention. When Koenig consulted him on the subject, Nicholson observed that "seventeen years before he had taken out a patent for machine printing, but he had abandoned it, thinking that it wouldn't do; and had never taken it up again." Indeed, the two machines were on different principles. Nor did Nicholson himself ever make any claim to priority of invention, when the success of Koenig's machine was publicly proclaimed by Mr. Walter of The Times some seven years later.

When Koenig, now settled abroad, heard of the attempts made in England to deny his merits as an inventor, he merely observed to his friend Bauer, "It is really too bad that these people, who have already robbed me of my invention, should now try to rob me of my reputation." Had he made any reply to the charges against him, it might have been comprised in a very few words: "When I arrived in England, no steam printing machine had ever before been seen; when I left it, the only printing machines in actual work were those which I had constructed." But Koenig never took the trouble to defend the originality of his invention in England, now that he had finally abandoned the field to others.

There can be no question as to the great improvements introduced in the printing machine by Mr. Applegath and Mr. Cowper; by Messrs. Hoe and Sons, of New York; and still later by the present Mr. Walter of The Times, which have brought the art of machine printing to an extraordinary degree of perfection and speed. But the original merits of an invention are not to be determined by a comparison of the first machine of the kind ever made with the last, after some sixty years' experience and skill have been applied in bringing it to perfection. Were the first condensing engine made at Soho—now to be seen at the Museum in South Kensington—in like manner to be compared with the last improved pumping-engine made yesterday, even the great James Watt might be made out to have been a very poor contriver. It would be much fairer to compare Koenig's steam-printing machine with the hand-press newspaper printing machine which it superseded. Though there were steam engines before Watt, and steamboats before Fulton, and steam locomotives before Stephenson, there were no steam printing presses before Koenig with which to compare them, Koenig's was undoubtedly the first, and stood unequalled and alone.

The rest of Koenig's life, after he retired to Germany, was spent in industry, if not in peace and quietness. He could not fail to be cast down by the utter failure of his English partnership, and the loss of the fruits of his ingenious labours. But instead of brooding over his troubles, he determined to break away from them, and begin the world anew. He was only forty-three when he left England, and he might yet be able to establish himself prosperously in life. He had his own head and hands to help him.

Though England was virtually closed against him, the whole continent of Europe was open to him, and presented a wide field for the sale of his printing machines.

While residing in England, Koenig had received many communications from influential printers in Germany. Johann Spencer and George Decker wrote to him in 1815, asking for particulars about his invention; but finding his machine too expensive,[7] the latter commissioned Koenig to send him a Stanhope printing press—the first ever introduced into Germany—the price of which was 95L. Koenig did this service for his friend, for although he stood by the superior merits of his own invention, he was sufficiently liberal to recognise the merits of the inventions of others. Now that he was about to settle in Germany, he was able to supply his friends and patrons on the spot.

The question arose, where was he to settle? He made enquiries about sites along the Rhine, the Neckar, and the Main. At last he was attracted by a specially interesting spot at Oberzell on the Main, near Wurzburg. It was an old disused convent of the Praemonstratensian monks. The place was conveniently situated for business, being nearly in the centre of Germany. The Bavarian Government, desirous of giving encouragement to so useful a genius, granted Koenig the use of the secularised monastery on easy terms; and there accordingly he began his operations in the course of the following year. Bauer soon joined him, with an order from Mr. Walter for an improved Times machine; and the two men entered into a partnership which lasted for life.

The partners had at first great difficulties to encounter in getting their establishment to work. Oberzell was a rural village, containing only common labourers, from whom they had to select their workmen. Every person taken into the concern had to be trained and educated to mechanical work by the partners themselves. With indescribable patience they taught these labourers the use of the hammer, the file, the turning-lathe, and other tools, which the greater number of them had never before seen, and of whose uses they were entirely ignorant. The machinery of the workshop was got together with equal difficulty piece by piece, some of the parts from a great distance,—the mechanical arts being then at a very low ebb in Germany, which was still suffering from the effects of the long continental war.

At length the workshop was fitted up, the old barn of the monastery being converted into an iron foundry.

Orders for printing machines were gradually obtained. The first came from Brockhaus, of Leipzig. By the end of the fourth year two other single-cylinder machines were completed and sent to Berlin, for use in the State printing office. By the end of the eighth year seven double-cylinder steam presses had been manufactured for the largest newspaper printers in Germany. The recognised excellence of Koenig and Bauer's book-printing machines—their perfect register, and the quality of the work they turned out—secured for them an increasing demand, and by the year 1829 the firm had manufactured fifty-one machines for the leading book printers throughout Germany. The Oberzell manufactory was now in full work, and gave regular employment to about 120 men.

A period of considerable depression followed. As was the case in England, the introduction of the printing machine in Germany excited considerable hostility among the pressmen. In some of the principal towns they entered into combinations to destroy them, and several printing machines were broken by violence and irretrievably injured. But progress could not be stopped; the printing machine had been fairly born, and must eventually do its work for mankind. These combinations, however, had an effect for a time. They deterred other printers from giving orders for the machines; and Koenig and Bauer were under the necessity of suspending their manufacture to a considerable extent. To keep their men employed, the partners proceeded to fit up a paper manufactory, Mr. Cotta, of Stuttgart, joining them in the adventure; and a mill was fitted up, embodying all the latest improvements in paper-making.

Koenig, however, did not live to enjoy the fruits or all his study, labour, toil, and anxiety; for, while this enterprise was still in progress, and before the machine trade had revived, he was taken ill, and confined to bed. He became sleepless; his nerves were unstrung; and no wonder. Brain disease carried him off on the 17th of January, 1833; and this good, ingenious, and admirable inventor was removed from all further care and trouble.

He died at the early age of fifty-eight, respected and beloved by all who knew him.

His partner Bauer survived to continue the business for twenty years longer. It was during this later period that the Oberzell manufactory enjoyed its greatest prosperity. The prejudices of the workmen gradually subsided when they found that machine printing, instead of abridging employment, as they feared it would do, enormously increased it; and orders accordingly flowed in from Berlin, Vienna, and all the leading towns and cities of Germany, Austria, Denmark, Russia, and Sweden. The six hundredth machine, turned out in 1847, was capable of printing 6000 impressions in the hour. In March, 1865, the thousandth machine was completed at Oberzell, on the occasion of the celebration of the fifty years' jubilee of the invention of the steam press by Koenig.

The sons of Koenig carried on the business; and in the biography by Goebel, it is stated that the manufactory of Oberzell has now turned out no fewer than 3000 printing machines. The greater number have been supplied to Germany; but 660 were sent to Russia, 61 to Asia, 12 to England, and 11 to America. The rest were despatched to Italy, Switzerland, Sweden, Spain, Holland, and other countries.

It remains to be said that Koenig and Bauer, united in life, were not divided by death. Bauer died on February 27, 1860, and the remains of the partners now lie side by side in the little cemetery at Oberzell, close to the scene of their labours and the valuable establishment which they founded.

Footnotes for Chapter VI.

[1] Koenig's letter in The Times, 8th December, 1814

[2] Koenig's letter in The Times, 8th December, 1814.

[3] Date of Patent, 29th April, 1790, No. 1748,

[4] Koenig's letter in The Times, 8th December, 1814.

[5] Mr. Richard Taylor, one of the partners in the patent, says, "Mr. Perry declined, alleging that he did not consider a newspaper worth so many years' purchase as would equal the cost of the machine."

[6] Mr. Richard Taylor, F.S.A., memoir in 'Philosophical Magazine' for October 1847, p. 300.

[7] The price of a single cylinder non-registering machine was advertised at 900L.; of a double ditto, 1400L.; and of a cylinder registering machine, 2000L.; added to which was 250L., 350L., and 500L. per annum for each of these machines so long as the patent lasted, or an agreed sum to be paid down at once.

CHAPTER VII.

THE WALTERS OF THE TIMES: INVENTION OF THE WALTER PRESS.

"Intellect and industry are never incompatible. There is more wisdom, and will be more benefit, in combining them than scholars like to believe, or than the common world imagine. Life has time enough for both, and its happiness will be increased by the union."—SHARON TURNER.

"I have beheld with most respect the man Who knew himself, and knew the ways before him, And from among them chose considerately, With a clear foresight, not a blindfold courage; And, having chosen, with a steadfast mind Pursued his purpose." HENRY TAYLOR—Philip van Artevelde.

The late John Walter, who adopted Koenig's steam printing press in printing The Times, was virtually the inventor of the modern newspaper. The first John Walter, his father, learnt the art of printing in the office of Dodsley, the proprietor of the 'Annual Register.' He afterwards pursued the profession of an underwriter, but his fortunes were literally shipwrecked by the capture of a fleet of merchantmen by a French squadron. Compelled by this loss to return to his trade, he succeeded in obtaining the publication of 'Lloyd's List,' as well as the printing of the Board of Customs. He also established himself as a publisher and bookseller at No. 8, Charing Cross. But his principal achievement was in founding The Times newspaper.

The Daily Universal Register was started on the 1st of January, 1785, and was described in the heading as "printed logographically." The type had still to be composed, letter by letter, each placed alongside of its predecessor by human fingers. Mr. Walter's invention consisted in using stereotyped words and parts of words instead of separate metal letters, by which a certain saving of time and labour was effected. The name of the 'Register' did not suit, there being many other publications bearing a similar title. Accordingly, it was re-named The Times, and the first number was issued from Printing House Square on the 1st of January, 1788.

The Times was at first a very meagre publication. It was not much bigger than a number of the old 'Penny Magazine,' containing a single short leader on some current topic, without any pretensions to excellence; some driblets of news spread out in large type; half a column of foreign intelligence, with a column of facetious paragraphs under the heading of "The Cuckoo;" while the rest of each number consisted of advertisements. Notwithstanding the comparative innocence of the contents of the early numbers of the paper, certain passages which appeared in it on two occasions subjected the publisher to imprisonment in Newgate. The extent of the offence, on one occasion, consisted in the publication of a short paragraph intimating that their Royal Highnesses the Prince of Wales and the Duke of York had "so demeaned themselves as to incur the just disapprobation of his Majesty!" For such slight offences were printers sent to gaol in those days.

Although the first Mr. Walter was a man of considerable business ability, his exertions were probably too much divided amongst a variety of pursuits to enable him to devote that exclusive attention to The Times which was necessary to ensure its success.

He possibly regarded it, as other publishers of newspapers then did, mainly as a means of obtaining a profitable business in job-printing. Hence, in the elder Walter's hands, the paper was not only unprofitable in itself, but its maintenance became a source of gradually increasing expenditure; and the proprietor seriously contemplated its discontinuance.

At this juncture, John Walter, junior, who had been taken into the business as a partner, entreated his father to entrust him with the sole conduct of the paper, and to give it "one more trial." This was at the beginning of 1803. The new editor and conductor was then only twenty-seven years of age. He had been trained to the manual work of a printer "at case," and passed through nearly every department in the office, literary and mechanical. But in the first place, he had received a very liberal education, first at Merchant Taylors' School, and afterwards at Trinity College, Oxford, where he pursued his classical studies with much success. He was thus a man of well-cultured mind; he had been thoroughly disciplined to work; he was, moreover, a man of tact and energy, full of expedients, and possessed by a passion for business. His father, urged by the young man's entreaties, at length consented, although not without misgivings, to resign into his hands the entire future control of The Times.

Young Walter proceeded forthwith to remodel the establishment, and to introduce improvements into every department, as far as the scanty capital at his command would admit.

Before he assumed the direction, The Times did not seek to guide opinion or to exercise political influence. It was a scanty newspaper—nothing more, Any political matters referred to were usually introduced in "Letters to the Editor," in the form in which Junius's Letters first appeared in the Public Advertiser. The comments on political affairs by the Editor were meagre and brief, and confined to a mere statement of supposed facts.

Mr. Walter, very much to the dismay of his father, struck out an entirely new course. He boldly stated his views on public affairs, bringing his strong and original judgment to bear upon the political and social topics of the day. He carefully watched and closely studied public opinion, and discussed general questions in all their bearings. He thus invented the modern Leading Article. The adoption of an independent line of politics necessarily led him to canvass freely, and occasionally to condemn, the measures of the Government. Thus, he had only been about a year in office as editor, when the Sidmouth Administration was succeeded by that of Mr. Pitt, under whom Lord Melville undertook the unfortunate Catamaran expedition. His Lordship's malpractices in the Navy Department had also been brought to light by the Commissioners of Naval Inquiry. On both these topics Mr. Walter spoke out freely in terms of reprobation; and the result was, that the printing for the Customs and the Government advertisements were at once removed from The Times office.

Two years later Mr. Pitt died, and an Administration succeeded which contained a portion of the political chiefs whom the editor had formerly supported on his undertaking the management of the paper. He was invited by one of them to state the injustice which had been done to him by the loss of the Customs printing, and a memorial to the Treasury was submitted for his signature, with a view to its recovery. But believing that the reparation of the injury in this manner was likely to be considered as a favour, entitling those who granted it to a certain degree of influence over the politics of the journal, Walter refused to sign it, or to have any concern in presenting the memorial. He did more; he wrote to those from whom the restoration of the employment was expected to come, disavowing all connection with the proceeding. The matter then dropped, and the Customs printing was never restored to the office.

This course was so unprecedented, and, as his father thought, was so very wrong-headed, that young Walter had for some time considerable difficulty in holding his ground and maintaining the independent position he had assumed. But with great tenacity of purpose he held on his course undismayed. He was a man who looked far ahead,—not so much taking into account the results at the end of each day or of each year, but how the plan he had laid down for conducting the paper would work out in the long run. And events proved that the high-minded course he had pursued with so much firmness of purpose was the wisest course after all.

Another feature in the management which showed clear-sightedness and business acuteness, was the pains which the Editor took to ensure greater celerity of information and dispatch in printing. The expense which he incurred in carrying out these objects excited the serious displeasure of his father, who regarded them as acts of juvenile folly and extravagance. Another circumstance strongly roused the old man's wrath. It appears that in those days the insertion of theatrical puffs formed a considerable source of newspaper income; and yet young Walter determined at once to abolish them. It is not a little remarkable that these earliest acts of Mr. Walter—which so clearly marked his enterprise and high-mindedness—should have been made the subject of painful comments in his father's will.

Notwithstanding this serious opposition from within, the power and influence of the paper visibly and rapidly grew. The new Editor concentrated in the columns of his paper a range of information such as had never before been attempted, or indeed thought possible. His vigilant eye was directed to every detail of his business. He greatly improved the reporting of public meetings, the money market, and other intelligence,—aiming at greater fulness and accuracy. In the department of criticism his labours were unwearied. He sought to elevate the character of the paper, and rendered it more dignified by insisting that it should be impartial. He thus conferred the greatest public service upon literature, the drama, and the fine arts, by protecting them against the evil influences of venal panegyric on the one hand, and of prejudiced hostility on the other.

But the most remarkable feature of The Times that which emphatically commended it to public support and ensured its commercial success—was its department of foreign intelligence. At the time that Walter undertook the management of the journal, Europe was a vast theatre of war; and in the conduct of commercial affairs—not to speak of political movements—it was of the most vital importance that early information should be obtained of affairs on the Continent. The Editor resolved to become himself the purveyor of foreign intelligence, and at great expense he despatched his agents in all directions, even in the track of armies; while others were employed, under various disguises and by means of sundry pretexts, in many parts of the Continent. These agents collected information, and despatched it to London, often at considerable risks, for publication in The Times, where it usually appeared long in advance of the government despatches.

The late Mr. Pryme, in his 'Autobiographic Recollections,' mentions a visit which he paid to Mr. Walter at his seat at Bearwood. "He described to me," says Mr. Pryme, "the cause of the large extension in the circulation of The Times. He was the first to establish a foreign correspondent. This was Henry Crabb Robinson, at a salary of 300L. a year.... Mr. Walter also established local reporters, instead of copying from the country papers. His father doubted the wisdom of such a large expenditure, but the son prophesied a gradual and certain success, which has actually been realised."

Mr. Robinson has described in his Diary the manner in which he became connected with the foreign correspondence. "In January, 1807," he says, "I received, through my friend J.D. Collier, a proposal from Mr. Walter that I should take up my residence at Altona, and become The Times correspondent. I was to receive from the editor of the 'Hamburger Correspondenten' all the public documents at his disposal, and was to have the benefit also of a mass of information, of which the restraints of the German Press did not permit him to avail himself. The honorarium I was to receive was ample with my habits of life. I gladly accepted the offer, and never repented having done so. My acquaintance with Mr. Walter ripened into friendship, and lasted as long as he lived."[1]

Mr. Robinson was forced to leave Germany by the Battle of Friedland and the Treaty of Tilsit, which resulted in the naval coalition against England. Returning to London, he became foreign editor of The Times until the following year, when he proceeded to Spain as foreign correspondent. Mr. Walter had also an agent in the track of the army in the unfortunate Walcheren expedition; and The Times announced the capitulation of Flushing forty-eight hours before the news had arrived by any other channel. By this prompt method of communicating public intelligence, the practice, which had previously existed, of systematically retarding the publication of foreign news by officials at the General Post Office, who made gain by selling them to the Lombard Street brokers, was effectually extinguished.

This circumstance, as well as the independent course which Mr. Walter adopted in the discussion of foreign politics, explains in some measure the opposition which he had to encounter in the transmission of his despatches. As early as the year 1805, when he had come into collision with the Government and lost the Customs printing, The Times despatches were regularly stopped at the outports, whilst those for the Ministerial journals were allowed to proceed. This might have crushed a weaker man, but it did not crush Walter. Of course he expostulated. He was informed at the Home Secretary's office that he might be permitted to receive his foreign papers as a favour. But as this implied the expectation of a favour from him in return, the proposal was rejected; and, determined not to be baffled, he employed special couriers, at great cost, for the purpose of obtaining the earliest transmission of foreign intelligence.

These important qualities—enterprise, energy, business tact, and public spirit—sufficiently account for his remarkable success. To these, however, must be added another of no small importance—discernment and knowledge of character. Though himself the head and front of his enterprise, it was necessary that he should secure the services and co-operation of men of first-rate ability; and in the selection of such men his judgment was almost unerring. By his discernment and munificence, he collected round him some of the ablest writers of the age. These were frequently revealed to him in the communications of correspondents—the author of the letters signed "Vetus" being thus selected to write in the leading columns of the Paper. But Walter himself was the soul of The Times. It was he who gave the tone to its articles, directed its influence, and superintended its entire conduct with unremitting vigilance.

Even in conducting the mechanical arrangements of the paper—a business of no small difficulty—he had often occasion to exercise promptness and boldness of decision in cases of emergency. Printers in those days were a rather refractory class of work men, and not unfrequently took advantage of their position to impose hard terms on their employers, especially in the daily press, where everything must be promptly done within a very limited time. Thus on one occasion, in 1810, the pressmen made a sudden demand upon the proprietor for an increase of wages, and insisted upon a uniform rate being paid to all hands, whether good or bad. Walter was at first disposed to make concessions to the men; but having been privately informed that a combination was already entered into by the compositors, as well as by the pressmen, to leave his employment suddenly, under circumstances that would have stopped the publication of the paper, and inflicted on him the most serious injury, he determined to run all risks, rather than submit to what now appeared to him in the light of an extortion.

The strike took place on a Saturday morning, when suddenly, and without notice, all the hands turned out. Mr. Walter had only a few hours' notice of it, but he had already resolved upon his course. He collected apprentices from half a dozen different quarters, and a few inferior workmen, who were glad to obtain employment on any terms. He himself stript to his shirt-sleeves, and went to work with the rest; and for the next six-and-thirty hours he was incessantly employed at case and at press. On the Monday morning, the conspirators, who had assembled to triumph over his ruin, to their inexpressible amazement saw The Times issue from the publishing office at the usual hour, affording a memorable example of what one man's resolute energy may accomplish in a moment of difficulty.

The journal continued to appear with regularity, though the printers employed at the office lived in a state of daily peril. The conspirators, finding themselves baffled, resolved upon trying another game. They contrived to have two of the men employed by Walter as compositors apprehended as deserters from the Royal Navy. The men were taken before the magistrate; but the charge was only sustained by the testimony of clumsy, perjured witnesses, and fell to the ground. The turn-outs next proceeded to assault the new hands, when Mr. Walter resolved to throw around them the protection of the law. By the advice of counsel, he had twenty-one of the conspirators apprehended and tried, and nineteen of them were found guilty and condemned to various periods of imprisonment. From that moment combination was at an end in Printing House Square.

Mr. Walter's greatest achievement was his successful application of steam power to newspaper printing. Although he had greatly improved the mechanical arrangements after he took command of the paper, the rate at which the copies could be printed off remained almost stationary. It took a very long time indeed to throw off, by the hand-labour of pressmen, the three or four thousand copies which then constituted the ordinary circulation of The Times. On the occasion of any event of great public interest being reported in the paper, it was found almost impossible to meet the demand for copies. Only about 300 copies could be printed in the hour, with one man to ink the types and another to work the press, while the labour was very severe. Thus it took a long time to get out the daily impression, and very often the evening papers were out before The Times had half supplied the demand.

Mr. Walter could not brook the tedium of this irksome and laborious process. To increase the number of impressions, he resorted to various expedients. The type was set up in duplicate, and even in triplicate; several Stanhope presses were kept constantly at work; and still the insatiable demands of the newsmen on certain occasions could not be met. Thus the question was early forced upon his consideration, whether he could not devise machinery for the purpose of expediting the production of newspapers. Instead of 300 impressions an hour, he wanted from 1500 to 2000. Although such a speed as this seemed quite as chimerical as propelling a ship through the water against wind and tide at fifteen miles an hour, or running a locomotive on a railway at fifty, yet Mr. Walter was impressed with the conviction that a much more rapid printing of newspapers was feasible than by the slow hand-labour process; and he endeavoured to induce several ingenious mechanical contrivers to take up and work out his idea.

The principle of producing impressions by means of a cylinder, and of inking the types by means of a roller, was not new. We have seen, in the preceding memoir, that as early as 1790 William Nicholson had patented such a method, but his scheme had never been brought into practical operation. Mr. Walter endeavoured to enlist Marc Isambard Brunel—one of the cleverest inventors of the day—in his proposed method of rapid printing by machinery; but after labouring over a variety of plans for a considerable time, Brunel finally gave up the printing machine, unable to make anything of it. Mr. Walter next tried Thomas Martyn, an ingenious young compositor, who had a scheme for a self-acting machine for working the printing press. He was supplied with the necessary funds to enable him to prosecute his idea; but Mr. Walter's father was opposed to the scheme, and when the funds became exhausted, this scheme also fell to the ground.

As years passed on, and the circulation of the paper increased, the necessity for some more expeditious method of printing became still more urgent. Although Mr. Walter had declined to enter into an arrangement with Bensley in 1809, before Koenig had completed his invention of printing by cylinders, it was different five years later, when Koenig's printing machine was actually at work. In the preceding memoir, the circumstances connected with the adoption of the invention by Mr. Walter are fully related; as well as the announcement made in The Times on the 29th of November, 1814—the day on which the first newspaper printed by steam was given to the world.

But Koenig's printing machine was but the beginning of a great new branch of industry. After he had left this country in disgust, it remained for others to perfect the invention; although the ingenious German was entitled to the greatest credit for having made the first satisfactory beginning. Great inventions are not brought forth at a heat. They are begun by one man, improved by another, and perfected by a whole host of mechanical inventors. Numerous patents were taken out for the mechanical improvement of printing. Donkin and Bacon contrived a machine in 1813, in which the types were placed on a revolving prism. One of them was made for the University of Cambridge, but it was found too complicated; the inking was defective; and the project was abandoned.

In 1816, Mr. Cowper obtained a patent (No.3974) entitled, "A Method of Printing Paper for Paper Hangings, and Other Purposes."

The principal feature of this invention consisted in the curving or bending of stereotype plates for the purpose of being printed in that form. A number of machines for printing in two colours, in exact register, was made for the Bank of England, and four millions of One Pound notes were printed before the Bank Directors determined to abolish their further issue. The regular mode of producing stereotype plates, from plaster of Paris moulds, took so much time, that they could not then be used for newspaper printing.

Two years later, in 1818, Mr. Cowper invented and patented (No. 4194) his great improvements in printing. It may be mentioned that he was then himself a printer, in partnership with Mr. Applegath, his brother-in-law. His invention consisted in the perfect distribution of the ink, by giving end motion to the rollers, so as to get a distribution crossways, as well as lengthways. This principle is at the very foundation of good printing, and has been adopted in every machine since made. The very first experiment proved that the principle was right. Mr. Cowper was asked by Mr. Walter to alter Koenig's machine at The Times office, so as to obtain good distribution. He adopted two of Nicholson's single cylinders and flat formes of type. Two "drums" were placed betwixt the cylinders to ensure accuracy in the register,—over and under which the sheet was conveyed in it s progress from one cylinder to the other,—the sheet being at all times firmly held between two tapes, which bound it to the cylinders and drums. This is commonly called, in the trade, a "perfecting machine;" that is, it printed the paper on both sides simultaneously, and is still much used for "book-work," whilst single cylinder machines are often used for provincial newspapers.

After this, Mr. Cowper designed the four cylinder machine for The Times,—by means of which from 4000 to 5000 sheets could be printed from one forme in the hour. In 1823, Mr. Applegath invented an improvement in the inking apparatus, by placing the distributing rollers at an angle across the distributing table, instead of forcing them endways by other means.

Mr. Walter continued to devote the same unremitting attention to his business as before. He looked into all the details, was familiar with every department, and, on an emergency, was willing to lend a hand in any work requiring more than ordinary despatch.

Thus, it is related of him that, in the spring of 1833, shortly after his return to Parliament as Member for Berkshire, he was at The Times office one day, when an express arrived from Paris, bringing the speech of the King of the French on the opening of the Chambers. The express arrived at 10 A.M., after the day's impression of the paper had been published, and the editors and compositors had left the office. It was important that the speech should be published at once; and Mr. Walter immediately set to work upon it. He first translated the document; then, assisted by one compositor, he took his place at the type-case, and set it up. To the amazement of one of the staff, who dropped in about noon, he "found Mr. Walter, M.P. for Berks, working in his shirt-sleeves!" The speech was set and printed, and the second edition was in the City by one o'clock. Had he not "turned to" as he did, the whole expense of the express service would have been lost. And it is probable that there was not another man in the whole establishment who could have performed the double work—intellectual and physical—which he that day executed with his own head and hands.

Such an incident curiously illustrates his eminent success in life. It was simply the result of persevering diligence, which shrank from no effort and neglected no detail; as well as of prudence allied to boldness, but certainly not "of chance;" and, above all, of highminded integrity and unimpeachable honesty. It is perhaps unnecessary to add more as to the merits of Mr. Walter as a man of enterprise in business, or as a public man and a Member of Parliament. The great work of his life was the development of his journal, the history of which forms the best monument to his merits and his powers.

The progressive improvement of steam printing machinery was not affected by Mr. Walter's death, which occurred in 1847. He had given it an impulse which it never lost. In 1846 Mr. Applegath patented certain important improvements in the steam press. The general disposition of his new machine was that of a vertical cylinder 200 inches in circumference, holding on it the type and distributing surfaces, and surrounded alternately by inking rollers and pressing cylinders. Mr. Applegath estimated in his specification that in his new vertical system the machine, with eight cylinders, would print about 10,000 sheets per hour. The new printing press came into use in 1848, and completely justified the anticipations of its projector.

Applegath's machine, though successfully employed at The Times office, did not come into general use. It was, to a large extent, superseded by the invention of Richard M. Hoe, of New York. Hoe's process consisted in placing the types upon a horizontal cylinder, against which the sheets were pressed by exterior and smaller cylinders. The types were arranged in segments of a circle, each segment forming a frame that could be fixed on the cylinder. These printing machines were made with from two to ten subsidiary cylinders. The first presses sent by Messrs. Hoe & Co. to this country were for Lloyd's Weekly Newspaper, and were of the six-cylinder size. These were followed by two ten-cylinder machines, ordered by the present Mr. Walter, for The Times. Other English newspaper proprietors—both in London and the provinces—were supplied with the machines, as many as thirty-five having been imported from America between 1856 and 1862. It may be mentioned that the two ten-cylinder Hoes made for The Times were driven at the rate of thirty-two revolutions per minute, which gives a printing rate of 19,200 per hour, or about 16,000 including stoppages.

Much of the ingenuity exercised both in the Applegath and Hoe Machines was directed to the "chase," which had to hold securely upon its curved face the mass of movable type required to form a page. And now the enterprise of the proprietor of The Times again came to the front. The change effected in the art of newspaper-printing, by the process of stereotypes, is scarcely inferior to that by which the late Mr. Walter applied steam-power to the printing press, and certainly equal to that by which the rotary press superseded the reciprocatory action of the flat machine.

Stereotyping has a curious history. Many attempts were made to obtain solid printing-surfaces by transfer from similar surfaces, composed, in the first place, of movable types. The first who really succeeded was one Ged, an Edinburgh goldsmith, who, after a series of difficult experiments, arrived at a knowledge of the art of stereotyping. The first method employed was to pour liquid stucco, of the consistency of cream, over the types; and this, when solid, gave a perfect mould. Into this the molten metal was poured, and a plate was produced, accurately resembling the page of type. As long ago as 1730, Ged obtained a privilege from the University of Cambridge for printing Bibles and Prayer-books after this method. But the workmen were dead against it, as they thought it would destroy their trade. The compositors and the pressmen purposely battered the letters in the absence of their employers. In consequence of this interference Ged was ruined, and died in poverty.

The art had, however, been born, and could not be kept down. It was revived in France, in Germany, and in America. Fifty years after the discovery of Ged, Tilloch and Foulis, of Glasgow, patented a similar invention, without knowing anything of what Ged had done; and after great labour and many experiments, they produced plates, the impressions from which could not be distinguished from those taken from the types from which they were cast. Some years afterwards, Lord Stanhope, to whom the art of printing is much indebted, greatly improved the art of stereotyping, though it was still quite inapplicable to newspaper printing. The merit of this latter invention is due to the enterprise of the present proprietor of The Times.

Mr. Walter began his experiments, aided by an ingenious Italian founder named Dellagana, early in 1856. It was ascertained that when papier-mache matrices were rapidly dried and placed in a mould, separate columns might be cast in them with stereotype metal, type high, planed flat, and finished with sufficient speed to get up the duplicate of a forme of four pages fitted for printing. Steps were taken to adapt these type-high columns to the Applegath Presses, then worked with polygonal chases. When the Hoe machines were introduced, instead of dealing with the separate columns, the papier-mache matrix was taken from the whole page at one operation, by roller-presses constructed for the purpose. The impression taken off in this manner is as perfect as if it had been made in the finest wax. The matrix is rapidly dried on heating surfaces, and then accurately adjusted in a casting machine curved to the exact circumference of the main drum of the printing press, and fitted with a terra-cotta top to secure a casting of uniform thickness. On pouring stereotype metal into this mould, a curved plate was obtained, which, after undergoing a certain amount of trimming at two machines, could be taken to press and set to work within twenty-five minutes from the time at which the process began.

Besides the great advantages obtained from uniform sets of the plates, which might be printed on different machines at the rate of 50,000 impressions an hour, or such additional number as might be required, there is this other great advantage, that there is no wear and tear of type in the curved chases by obstructive friction; and that the fount, instead of wearing out in two years, might last for twenty; for the plates, after doing their work for one day, are melted down into a new impression for the next day's printing. At the same time, the original type-page, safe from injury, can be made to yield any number of copies that may be required by the exigencies of the circulation. It will be sufficiently obvious that by the multiplication of stereotype plates and printing machines, there is practically no limit to the number of copies of a newspaper that may be printed within the time which the process now usually occupies.

This new method of newspaper stereotyping was originally employed on the cylinders of the Applegath and Hoe Presses. But it is equally applicable to those of the Walter Press, a brief description of which we now subjoin. As the construction of the first steam newspaper machine was due to the enterprise of the late Mr. Walter, so the construction of this last and most improved machine is due in like manner to the enterprise of his son. The new Walter Press is not, like Applegath and Cowper's, and Hoe's, the improvement of an existing arrangement, but an almost entirely original invention.

In the Reports of the Jurors on the "Plate, Letterpress, and other modes of Printing," at the International Exhibition of 1862, the following passage occurs:—"It is incumbent on the reporters to point out that, excellent and surprising as are the results achieved by the Hoe and Applegath Machines, they cannot be considered satisfactory while those machines themselves are so liable to stoppages in working. No true mechanic can contrast the immense American ten-cylinder presses of The Times with the simple calico-printing machine, without feeling that the latter furnishes the true type to which the mechanism for newspaper printing should as much as possible approximate."

On this principle, so clearly put forward, the Inventors of the Walter Press proceeded in the contrivance of the new machine. It is true that William Nicholson, in his patent of 1790, prefigured the possibility of printing on "paper, linen, cotton, woollen, and other articles," by means of type fixed on the outer surface of a revolving cylinder; but no steps were taken to carry his views into effect. Sir Rowland Hill also, before he became connected with Post Office reform, revived the contrivance of Nicholson, and referred to it in his patent of 1835 (No. 6762); and he also proposed to use continuous rolls of paper, which Fourdrinier and Donkin had made practicable by their invention of the paper-making machine about the year 1804; but both Nicholson's and Hill's patents remained a dead letter.[2]

It may be easy to conceive a printing machine, or even to make a model of one; but to construct an actual working printing press, that must be sure and unfailing in its operations, is a matter surrounded with difficulties. At every step fresh contrivances have to be introduced; they have to be tried again and again; perhaps they are eventually thrown aside to give place to new arrangements. Thus the head of the inventor is kept in a state of constant turmoil. Sometimes the whole machine has to be remodelled from beginning to end. One step is gained by degrees, then another; and at last, after years of labour, the new invention comes before the world in the form of a practical working machine.

In 1862 Mr. Walter began in The Times office, with tools and machinery of his own, experiments for constructing a perfecting press which should print the paper from rolls of paper instead of from sheets. Like his father, Mr. Walter possessed an excellent discrimination of character, and selected the best men to aid him in his important undertaking. Numerous difficulties had, of course, to be surmounted. Plans were varied from time to time; new methods were tried, altered, and improved, simplification being aimed at throughout. Six long years passed in this pursuit of the possible. At length the clear light dawned. In 1868 Mr. Walter ventured to order the construction of three machines on the pattern of the first complete one which had been made. By the end of 1869 these were finished and placed in a room by themselves; and a fourth was afterwards added. There the printing of The Times is now done, in less than half the time it previously occupied, and with one-fifth the number of hands.

The most remarkable feature in the Walter Press is its wonderful simplicity of construction. Simplicity of arrangement is always the beau ideal of the mechanical engineer. This printing press is not only simple, but accurate, compact, rapid, and economical.

While each of the ten-feeder Hoe Machines occupies a large and lofty room, and requires eighteen men to feed and work it, the new Walter Machine occupies a space of only about 14 feet by 5, or less than any newspaper machine yet introduced; and it requires only three lads to take away, with half the attention of an overseer, who easily superintends two of the machines while at work. The Hoe Machine turns out 7000 impressions printed on both sides in the hour, whereas the Walter Machine turns out 12,000 impressions completed in the same time.

The new Walter Press does not in the least resemble any existing printing machine, unless it be the calendering machine which furnished its type. At the printing end it looks like a collection of small cylinders or rollers. The first thing to be observed is the continuous roll of paper four miles long, tightly mounted on a reel, which, when the machine is going, flies round with immense rapidity. The web of paper taken up by the first roller is led into a series of small hollow cylinders filled with water and steam, perforated with thousands of minute holes. By this means the paper is properly damped before the process of printing is begun. The roll of paper, drawn by nipping rollers, next flies through to the cylinder on which the stereotype plates are fixed, so as to form the four pages of the ordinary sheet of The Times; there it is lightly pressed against the type and printed; then it passes downwards round another cylinder covered with cloth, and reversed; next to the second type-covered roller, where it takes the impression exactly on the other side of the remaining four pages. It next reaches one of the most ingenious contrivances of the invention—the cutting machinery, by means of which the paper is divided by a quick knife into the 5500 sheets of which the entire web consists. The tapes hurry the now completely printed newspaper up an inclined plane, from which the divided sheets are showered down in a continuous stream by an oscillating frame, where they are met by two boys, who adjust the sheets as they fall. The reel of four miles long is printed and divided into newspapers complete in about twenty-five minutes.

The machine is almost entirely self-acting, from the pumping-up of the ink into the ink-box out of the cistern below stairs, to the registering of the numbers as they are printed in the manager's room above. It is always difficult to describe a machine in words. Nothing but a series of sections and diagrams could give the reader an idea of the construction of this unrivalled instrument. The time to see it and wonder at it is when the press is in full work. And even then you can see but little of its construction, for the cylinders are wheeling round with immense velocity. The rapidity with which the machine works may be inferred from the fact that the printing cylinders (round which the stereotyped plates are fixed), while making their impressions on the paper, travel at the surprising speed of 200 revolutions a minute, or at the rate of about nine miles an hour!

Contrast this speed with the former slowness. Go back to the beginning of the century. Before the year 1814 the turn-out of newspapers was only about 300 single impressions in an hour—that is, impressions printed on only one side of the paper. Koenig by his invention increased the issue to 1100 impressions. Applegath and Cowper by their four-cylinder machine increased the issue to 4000, and by the eight-cylinder machine to 10,000 an hour. But these were only impressions printed on one side of the paper. The first perfecting press—that is, printing simultaneously the paper on both sides—was the Walter, the speed of which has been raised to 12,000, though, if necessary, it can produce excellent work at the rate of 17,000 complete copies of an eight-page paper per hour. Then, with the new method of stereotyping—by means of which the plates can be infinitely multiplied and by the aid of additional machines, the supply of additional impressions is absolutely unlimited.

The Walter Press is not a monopoly. It is manufactured at The Times office, and is supplied to all comers. Among the other daily papers printed by its means in this country are the Daily News, the Scotsmam, and the Birmingham Daily Post. The first Walter Press was sent to America in 1872, where it was employed to print the Missouri Republican at St. Louis, the leading newspaper of the Mississippi Valley. An engineer and a skilled workman from The Times office accompanied the machinery. On arriving at St. Louis—the materials were unpacked, lowered into the machine-room, where they were erected and ready for work in the short space of five days.

The Walter Press was an object of great interest at the Centennial Exhibition held at Philadelphia in 1876, where it was shown printing the New Fork Times one of the most influential journals in America. The press was surrounded with crowds of visitors intently watching its perfect and regular action, "like a thing of life." The New York Times said of it: "The Walter Press is the most perfect printing press yet known to man; invented by the most powerful journal of the Old World, and adopted as the very best press to be had for its purposes by the most influential journal of the New World.... It is an honour to Great Britain to have such an exhibit in her display, and a lasting benefit to the printing business, especially to newspapers.... The first printing press run by steam was erected in the year 1814 in the office of The Times by the father of him who is the present proprietor of that world-famous journal. The machine of 1814 was described in The Times of the 29th November in that year, and the account given of it closed in these words: 'The whole of these complicated acts is performed with such a velocity and simultaneosness of movement that no less than 1100 sheets are impressed in one hour.' Mirabile dictu! And the Walter Press of to-day can run off 17,000 copies an hour printed on both sides. This is not bad work for one man's lifetime."

It is unnecessary to say more about this marvellous machine. Its completion forms the crown of the industry which it represents, and of the enterprise of the journal which it prints.

Footnotes for Chapter VII.

[1] Diary, Reminiscences, and Correspondence of Henry Crabb Robinson, Barrister-at-Law, F.S.A., i. 231.

[2] After the appearance of my article on the Koenig and Walter Presses in Macmillan's Magazine for December, 1869, I received the following letter from Sir Rowland Hill:—

"Hampstead" January 5th, 1870.

"My dear sir,

"In your very interesting article in Macmillan's Magazine on the subject of the printing machine, you have unconsciously done me some injustice. To convince yourself of this, you have only to read the enclosed paper. The case, however, will be strengthened when I tell you that as far back as the year 1856, that is, seven years after the expiry of my patent, I pointed out to Mr. Mowbray Morris, the manager of The Times, the fitness of my machine for the printing of that journal, and the fact that serious difficulties to its adoption had been removed. I also, at his request, furnished him with a copy of the document with which I now trouble you. Feeling sure that you would like to know the truth on any subject of which you may treat, I should be glad to explain the matter more fully, and for this purpose will, with your permission, call upon you at any time you may do me the favour to appoint. "Faithfully yours,

"Rowland Hill."

On further enquiry I obtained the Patent No. 6762; but found that nothing practical had ever come of it. The pamphlet enclosed by Sir Rowland Hill in the above letter is entitled 'The Rotary Printing Machine.' It is very clever and ingenious, like everything he did. But it was still left for some one else to work out the invention into a practical working printing-press. The subject is fully referred to in the 'Life of Sir Rowland Hill' (i. 224,525). In his final word on the subject, Sir Rowland "gladly admits the enormous difficulty of bringing a complex machine into practical use," a difficulty, he says, which "has been most successfully overcome by the patentees of the Walter Press."

CHAPTER VIII.

WILLIAM CLOWES: INTRODUCER OF BOOK-PRINTING BY STEAM.

"The Images of men's wits and knowledges remain in Books, exempted from the wrong of time, and capable of perpetual renovation. Neither are they fitly to be called Images, because they generate still, and cast their seeds in the minds of others, provoking and causing infinite actions and opinions in succeeding ages; so that, if the invention of the Ship was thought so noble, which carrieth riches and commodities from place to place, and consociateth the most remote Regions in participation of their Fruits, how much more are letters to be magnified, which, as Ships, pass through the vast Seas of time, and make ages so distant to participate of the wisdom, illuminations, and inventions, the one of the other?"—Bacon, On the Proficience and Advancement of Learning.

Steam has proved as useful and potent in the printing of books as in the printing of newspapers. Down to the end of last century, "the divine art," as printing was called, had made comparatively little progress. That is to say, although books could be beautifully printed by hand labour, they could not be turned out in any large numbers.

The early printing press was rude. It consisted of a table, along which the forme of type, furnished with a tympan and frisket, was pushed by hand. The platen worked vertically between standards, and was brought down for the impression, and raised after it, by a common screw, worked by a bar handle. The inking was performed by balls covered with skin pelts; they were blacked with ink, and beaten down on the type by the pressman. The inking was consequently irregular.

In 1798, Earl Stanhope perfected the press that bears his name. He did not patent it, but made his invention over to the public. In 1818, Mr. Cowper greatly improved the inking of formes used in the Stanhope and other presses, by the use of a hand roller covered with a composition of glue and treacle, in combination with a distributing table. The ink was thus applied in a more even manner, and with a considerable decrease of labour. With the Stanhope Press, printing was as far advanced as it could possibly be by means of hand labour. About 250 impressions could be taken off, on one side, in an hour.

But this, after all, was a very small result. When books could be produced so slowly, there could be no popular literature. Books were still articles for the few, instead of for the many. Steam power, however, completely altered the state of affairs. When Koenig invented his steam press, he showed by the printing of Clarkson's 'Life of Penn'—the first sheets ever printed with a cylindrical press—that books might be printed neatly, as well as cheaply, by the new machine. Mr. Bensley continued the process, after Koenig left England; and in 1824, according to Johnson in his 'Typographia,' his son was "driving an extensive business."

In the following year, 1825, Archibald Constable, of Edinburgh, propounded his plan for revolutionising the art of bookselling. Instead of books being articles of luxury, he proposed to bring them into general consumption. He would sell them, not by thousands, but by hundreds of thousands, "ay, by millions;" and he would accomplish this by the new methods of multiplication—by machine printing and by steam power. Mr. Constable accordingly issued a library of excellent books; and, although he was ruined—not by this enterprise, but the other speculations into which he entered—he set the example which other enterprising minds were ready to follow. Amongst these was Charles Knight, who set the steam presses of William Clowes to work, for the purposes of the Society for the Diffusion of Useful Knowledge.

William Clowes was the founder of the vast printing establishment from which these sheets are issued; and his career furnishes another striking illustration of the force of industry and character. He was born on the 1st of January, 1779. His father was educated at Oxford, and kept a large school at Chichester; but dying when William was but an infant, he left his widow, with straitened means, to bring up her family. At a proper age William was bound apprentice to a printer at Chichester; and, after serving him for seven years, he came up to London, at the beginning of 1802, to seek employment as a journeyman. He succeeded in finding work at a small office on Tower Hill, at a small wage. The first lodgings he took cost him 5s. a week; but finding this beyond his means he hired a room in a garret at 2s. 6d., which was as much as he could afford out of his scanty earnings.

The first job he was put to, was the setting-up of a large poster-bill—a kind of work which he had been accustomed to execute in the country; and he knocked it together so expertly that his master, Mr. Teape, on seeing what he could do, said to him, "Ah! I find you are just the fellow for me." The young man, however, felt so strange in London, where he was without a friend or acquaintance, that at the end of the first month he thought of leaving it; and yearned to go back to his native city. But he had not funds enough to enable him to follow his inclinations, and he accordingly remained in the great City, to work, to persevere, and finally to prosper. He continued at Teape's for about two years, living frugally, and even contriving to save a little money.

He then thought of beginning business on his own account. The small scale on which printing was carried on in those days enabled him to make a start with comparatively little capital. By means of his own savings and the help of his friends, he was enabled to take a little printing-office in Villiers Street, Strand, about the end of 1803; and there he began with one printing press, and one assistant. His stock of type was so small, that he was under the necessity of working it from day to day like a banker's gold. When his first job came in, he continued to work for the greater part of three nights, setting the type during the day, and working it off at night, in order that the type might be distributed for resetting on the following morning. He succeeded, however, in executing his first job to the entire satisfaction of his first customer.

His business gradually increased, and then, with his constantly saved means, he was enabled to increase his stock of type, and to undertake larger jobs. Industry always tells, and in the long-run leads to prosperity. He married early, but he married well. He was only twenty-four when he found his best fortune in a good, affectionate wife. Through this lady's cousin, Mr. Winchester, the young printer was shortly introduced to important official business. His punctual execution of orders, the accuracy of his work, and the despatch with which he turned it out soon brought him friends, and his obliging and kindly disposition firmly secured them. Thus, in a few years, the humble beginner with one press became a printer on a large scale.

The small concern expanded into a considerable printing-office in Northumberland Court, which was furnished with many presses and a large stock of type. The office was, unfortunately, burnt down; but a larger office rose in its place.

What Mr. Clowes principally aimed at, in carrying on his business, was accuracy, speed, and quantity. He did not seek to produce editions de luxe in limited numbers, but large impressions of works in popular demand—travels, biographies, histories, blue-books, and official reports, in any quantity. For this purpose, he found the process of hand-printing too tedious, as well as too costly; and hence he early turned his attention to book printing by machine presses, driven by steam power,—in this matter following the example of Mr. Walter of the Times, who had for some years employed the same method for newspaper printing.

Applegath & Cowper's machines had greatly advanced the art of printing. They secured perfect inking and register; and the sheets were printed off more neatly, regularly, and expeditiously; and larger sheets could be printed on both sides, than by any other method. In 1823, accordingly, Mr. Clowes erected his first steam presses, and he soon found abundance of work for them. But to produce steam requires boilers and engines, the working of which occasions smoke and noise. Now, as the printing-office, with its steam presses, was situated in Northumberland Court, close to the palace of the Duke of Northumberland, at Charing Cross, Mr. Clowes was required to abate the nuisance, and to stop the noise and dirt occasioned by the use of his engines. This he failed to do, and the Duke commenced an action against him.

The case was tried in June, 1824, in the Court of Common Pleas. It was ludicrous to hear the extravagant terms in which the counsel for the plaintiff and his witnesses described the nuisance—the noise made by the engine in the underground cellar, some times like thunder, at other times like a thrashing-machine, and then again like the rumbling of carts and waggons. The printer had retained the Attorney-general, Mr. Copley, afterwards Lord Lyndhurst, who conducted his case with surpassing ability. The cross-examination of a foreign artist, employed by the Duke to repaint some portraits of the Cornaro family by Titian, is said to have been one of the finest things on record. The sly and pungent humour, and the banter with which the counsel derided and laughed down this witness, were inimitable. The printer won his case; but he eventually consented to remove his steam presses from the neighbourhood, on the Duke paying him a certain sum to be determined by the award of arbitrators.

It happened, about this period, that a sort of murrain fell upon the London publishers. After the failure of Constable at Edinburgh, they came down one after another, like a pack of cards.

Authors are not the only people who lose labour and money by publishers; there are also cases where publishers are ruined by authors. Printers also now lost heavily. In one week, Mr. Clowes sustained losses through the failure of London publishers to the extent of about 25,000L. Happily, the large sum which the arbitrators awarded him for the removal of his printing presses enabled him to tide over the difficulty; he stood his ground unshaken, and his character in the trade stood higher than ever.

In the following year Mr. Clowes removed to Duke Street, Blackfriars, to premises until then occupied by Mr. Applegath, as a printer; and much more extensive buildings and offices were now erected. There his business transactions assumed a form of unprecedented magnitude, and kept pace with the great demand for popular information which set in with such force about fifty years ago. In the course of ten years—as we find from the 'Encyclopaedia Metropolitana'—there were twenty of Applegath & Cowper's machines, worked by two five-horse engines. From these presses were issued the numerous admirable volumes and publications of the Society for the Diffusion of Useful Knowledge; the treatises on 'Physiology,' by Roget, and 'Animal Mechanics,' by Charles Bell; the 'Elements of Physics,' by Neill Arnott; 'The Pursuit of Knowledge under Difficulties,' by G. L. Craik, a most fascinating book; the Library of Useful Knowledge; the 'Penny Magazine,' the first illustrated publication; and the 'Penny Cyclopaedia,' that admirable compendium of knowledge and science.

These publications were of great value. Some of them were printed in unusual numbers. The 'Penny Magazine,' of which Charles Knight was editor, was perhaps too good, because it was too scientific. Nevertheless, it reached a circulation of 200,000 copies. The 'Penny Cyclopaedia' was still better. It was original, and yet cheap. The articles were written by the best men that could be found in their special departments of knowledge. The sale was originally 75,000 weekly; but, as the plan enlarged, the price was increased from 1d. to 2d., and then to 4d. At the end of the second year, the circulation had fallen to 44,000; and at the end of the third year, to 20,000.

It was unfortunate for Mr. Knight to be so much under the influence of his Society. Had the Cyclopaedia been under his own superintendence, it would have founded his fortune. As it was, he lost over 30,000L. by the venture. The 'Penny Magazine' also went down in circulation, until it became a non-paying publication, and then it was discontinued. It is curious to contrast the fortunes of William Chambers of Edinburgh with those of Charles Knight of London. 'Chambers's Edinburgh Journal' was begun in February, 1832, and the 'Penny Magazine' in March, 1832.

Chambers was perhaps shrewder than Knight. His journal was as good, though without illustrations; but he contrived to mix up amusement with useful knowledge. It may be a weakness, but the public like to be entertained, even while they are feeding upon better food. Hence Chambers succeeded, while Knight failed. The 'Penny Magazine' was discontinued in 1845, whereas 'Chambers's Edinburgh Journal' has maintained its popularity to the present day. Chambers, also, like Knight, published an 'Encyclopaedia,' which secured a large circulation. But he was not trammelled by a Society, and the 'Encyclopaedia' has become a valuable property.

The publication of these various works would not have been possible without the aid of the steam printing press. When Mr. Edward Cowper was examined before a Committee of the House of Commons, he said, "The ease with which the principles and illustrations of Art might be diffused is, I think, so obvious that it is hardly necessary to say a word about it. Here you may see it exemplified in the 'Penny Magazine.' Such works as this could not have existed without the printing machine." He was asked, "In fact, the mechanic and the peasant, in the most remote parts of the country, have now an opportunity of seeing tolerably correct outlines of form which they never could behold before?" To which he answered, "Exactly; and literally at the price they used to give for a song." "Is there not, therefore, a greater chance of calling genius into activity?" "Yes," he said, "not merely by books creating an artist here and there, but by the general elevation of the taste of the public."

Mr. Clowes was always willing to promote deserving persons in his office. One of these rose from step to step, and eventually became one of the most prosperous publishers in London. He entered the service as an errand-boy, and got his meals in the kitchen. Being fond of reading, he petitioned Mrs. Clowes to let him sit somewhere, apart from the other servants, where he might read his book in quiet. Mrs. Clowes at length entreated her husband to take him into the office, for "Johnnie Parker was such a good boy." He consented, and the boy took his place at a clerk's desk. He was well-behaved, diligent, and attentive. As he advanced in years, his steady and steadfast conduct showed that he could be trusted. Young fellows like this always make their way in life; for character invariably tells, not only in securing respect, but in commanding confidence. Parker was promoted from one post to another, until he was at length appointed overseer over the entire establishment.

A circumstance shortly after occurred which enabled Mr. Clowes to advance him, though greatly to his own inconvenience, to another important post. The Syndics of Cambridge were desirous that Mr. Clowes should go down there to set their printing-office in order; they offered him 400L. a year if he would only appear occasionally, and see that the organisation was kept complete. He declined, because the magnitude of his own operations had now become so great that they required his unremitting attention. He, however strongly recommended Parker to the office, though he could ill spare him. But he would not stand in the young man's way, and he was appointed accordingly. He did his work most effectually at Cambridge, and put the University Press into thorough working order.

As the 'Penny Magazine' and other publications of the Society of Useful Knowledge were now making their appearance, the clergy became desirous of bringing out a religious publication of a popular character, and they were in search for a publisher. Parker, who was well known at Cambridge, was mentioned to the Bishop of London as the most likely person. An introduction took place, and after an hour's conversation with Parker, the Bishop went to his friends and said, "This is the very man we want." An offer was accordingly made to him to undertake the publication of the 'Saturday Magazine' and the other publications of the Christian Knowledge Society, which he accepted. It is unnecessary to follow his fortunes. His progress was steady; he eventually became the publisher of 'Fraser's Magazine' and of the works of John Stuart Mill and other well-known writers. Mill never forgot his appreciation and generosity; for when his 'System of Logic' had been refused by the leading London publishers, Parker prized the book at its rightful value and introduced it to the public.

To return to Mr. Clowes. In the course of a few years, the original humble establishment of the Sussex compositor, beginning with one press and one assistant, grew up to be one of the largest printing-offices in the world. It had twenty-five steam presses, twenty-eight hand-presses, six hydraulic presses, and gave direct employment to over five hundred persons, and indirect employment to probably more than ten times that number. Besides the works connected with his printing-office, Mr. Clowes found it necessary to cast his own types, to enable him to command on emergency any quantity; and to this he afterwards added stereotyping on an immense scale. He possessed the power of supplying his compositors with a stream of new type at the rate of about 50,000 pieces a day. In this way, the weight of type in ordinary use became very great; it amounted to not less than 500 tons, and the stereotyped plates to about 2500 tons the value of the latter being not less than half a million sterling.

Mr. Clowes would not hesitate, in the height of his career, to have tons of type locked up for months in some ponderous blue-book. To print a report of a hundred folio pages in the course of a day or during a night, or of a thousand pages in a week, was no uncommon occurrence. From his gigantic establishment were turned out not fewer than 725,000 printed sheets, or equal to 30,000 volumes a week. Nearly 45,000 pounds of paper were printed weekly. The quantity printed on both sides per week, if laid down in a path of 22 1/4 inches broad, would extend 263 miles in length.

About the year 1840, a Polish inventor brought out a composing machine, and submitted it to Mr. Clowes for approval. But Mr. Clowes was getting too old to take up and push any new invention.

He was also averse to doing anything to injure the compositors, having once been a member of the craft. At the same time he said to his son George, "If you find this to be a likely machine, let me know. Of course we must go with the age. If I had not started the steam press when I did, where should I have been now?" On the whole, the composing machine, though ingenious, was incomplete, and did not come into use at that time, nor indeed for a long time after. Still, the idea had been born, and, like other inventions, became eventually developed into a useful working machine. Composing machines are now in use in many printing-offices, and the present Clowes' firm possesses several of them. Those in The Times newspaper office are perhaps the most perfect of all.

Mr. Clowes was necessarily a man of great ability, industry, and energy. Whatever could be done in printing, that he would do. He would never admit the force of any difficulty that might be suggested to his plans. When he found a person ready to offer objections, he would say, "Ah! I see you are a difficulty-maker: you will never do for me."

Mr. Clowes died in 1847, at the age of sixty-eight. There still remain a few who can recall to mind the giant figure, the kindly countenance, and the gentle bearing of this "Prince of Printers," as he was styled by the members of his craft. His life was full of hard and useful work; and it will probably be admitted that, as the greatest multiplier of books in his day, and as one of the most effective practical labourers for the diffusion of useful knowledge, his name is entitled to be permanently associated, not only with the industrial, but also with the intellectual development of our time.

CHAPTER IX.

CHARLES BIANCONI: A LESSON OF SELF-HELP IN IRELAND.

"I beg you to occupy yourself in collecting biographical notices respecting the Italians who have honestly enriched themselves in other regions, particularly referring to the obstacles of their previous life, and to the efforts and the means which they employed for vanquishing them, as well as to the advantages which they secured for themselves, for the countries in which they settled, and for the country to which they owed their birth."—GENERAL MENABREA, Circular to Italian Consuls.

When Count Menabrea was Prime Minister of Italy, he caused a despatch to be prepared and issued to Italian Consuls in all parts of the world, inviting them to collect and forward to him "biographical notices respecting the Italians who have honourably advanced themselves in foreign countries."

His object, in issuing the despatch, was to collect information as to the lives of his compatriots living abroad, in order to bring out a book similar to 'Self-help,' the examples cited in which were to be drawn exclusively from the lives of Italian citizens. Such a work, he intimated, "if it were once circulated among the masses, could not fail to excite their emulation and encourage them to follow the examples therein set forth," while "in the course of time it might exercise a powerful influence on the increased greatness of our country."

We are informed by Count Menabrea that, although no special work has been published from the biographical notices collected in answer to his despatch, yet that the Volere e Potere ('Will is

Power') of Professor Lessona, issued a few years ago, sufficiently answers the purpose which he contemplated, and furnishes many examples of the patient industry and untiring perseverance of Italians in all parts of the world. Many important illustrations of life and character are necessarily omitted from Professor Lessona's interesting work. Among these may be mentioned the subject of the following pages,—a distinguished Italian who entirely corresponds to Count Menabrea's description—one who, in the face of the greatest difficulties, raised himself to an eminent public position, at the same time that he conferred the greatest benefits upon the country in which he settled and carried on his industrial operations. We mean Charles Bianconi, and his establishment of the great system of car communication through out Ireland.[1]

Charles Bianconi was born in 1786, at the village of Tregolo, situated in the Lombard Highlands of La Brianza, about ten miles from Como. The last elevations of the Alps disappear in the district; and the great plain of Lombardy extends towards the south. The region is known for its richness and beauty; the inhabitants being celebrated for the cultivation of the mulberry and the rearing of the silkworm, the finest silk in Lombardy being produced in the neighbourhood. Indeed, Bianconi's family, like most of the villagers, maintained themselves by the silk culture.

Charles had three brothers and one sister. When of a sufficient age, he was sent to school. The Abbe Radicali had turned out some good scholars; but with Charles Bianconi his failure was complete. The new pupil proved a tremendous dunce. He was very wild, very bold, and very plucky; but he learned next to nothing.

Learning took as little effect upon him as pouring water upon a duck's back. Accordingly, when he left school at the age of sixteen, he was almost as ignorant as when he had entered it; and a great deal more wilful.

Young Bianconi had now arrived at the age at which he was expected to do something for his own maintenance. His father wished to throw him upon his own resources; and as he would soon be subject to the conscription, he thought of sending him to some foreign country in order to avoid the forced service. Young fellows, who had any love of labour or promptings of independence in them, were then accustomed to leave home and carry on their occupations abroad. It was a common practice for workmen in the neighbourhood of Como to emigrate to England and carry on various trades; more particularly the manufacture and sale of barometers, looking-glasses, images, prints, pictures, and other articles.

Accordingly, Bianconi's father arranged with one Andrea Faroni to take the young man to England and instruct him in the trade of print-selling. Bianconi was to be Faroni's apprentice for eighteen months; and in the event of his not liking the occupation, he was to be placed under the care of Colnaghi, a friend of his father's, who was then making considerable progress as a print-seller in London; and who afterwards succeeded in achieving a considerable fortune and reputation.

Bianconi made his preparations for leaving home. A little festive entertainment was given at a little inn in Como, at which the whole family were present. It was a sad thing for Bianconi's mother to take leave of her boy, wild though he was. On the occasion of this parting ceremony, she fainted outright, at which the young fellow thought that things were assuming a rather serious aspect. As he finally left the family home at Tregolo, the last words his mother said to him were these—words which he never forgot: "When you remember me, think of me as waiting at this window, watching for your return."

Besides Charles Bianconi, Faroni took three other boys under his charge. One was the son of a small village innkeeper, another the son of a tailor, and the third the son of a flax-dealer. This party, under charge of the Padre, ascended the Alps by the Val San Giacomo road. From the summit of the pass they saw the plains of Lombardy stretching away in the blue distance. They soon crossed the Swiss frontier, and then Bianconi found himself finally separated from home. He now felt, that without further help from friends or relatives, he had his own way to make in the world.

The party of travellers duly reached England; but Faroni, without stopping in London, took them over to Ireland at once. They reached Dublin in the summer of 1802, and lodged in Temple Bar, near Essex Bridge. It was some little time before Faroni could send out the boys to sell pictures. First he had the leaden frames to cast; then they had to be trimmed and coloured; and then the pictures—mostly of sacred subjects, or of public characters—had to be mounted. The flowers; which were of wax, had also to be prepared and finished, ready for sale to the passers-by.

When Bianconi went into the streets of Dublin to sell his mounted prints, he could not speak a word of English. He could only say, "Buy, buy!" Everybody spoke to him an unknown tongue. When asked the price, he could only indicate by his fingers the number of pence he wanted for his goods. At length he learned a little English,—at least sufficient "for the road;" and then he was sent into the country to sell his merchandize. He was despatched every Monday morning with about forty shillings' worth of stock, and ordered to return home on Saturdays, or as much sooner as he liked, if he had sold all the pictures. The only money his master allowed him at starting was fourpence. When Bianconi remonstrated at the smallness of the amount, Faroni answered, "While you have goods you have money; make haste to sell your goods!"

During his apprenticeship, Bianconi learnt much of the country through which he travelled. He was constantly making acquaintances with new people, and visiting new places. At Waterford he did a good trade in small prints. Besides the Scripture pieces, he sold portraits of the Royal Family, as well as of Bonaparte and his most distinguished generals. "Bony" was the dread of all magistrates, especially in Ireland. At Passage, near Waterford, Bianconi was arrested for having sold a leaden framed picture of the famous French Emperor. He was thrown into a cold guard-room, and spent the night there without bed, or fire, or food. Next morning he was discharged by the magistrate, but cautioned that he must not sell any more of such pictures.

Many things struck Bianconi in making his first journeys through Ireland. He was astonished at the dram-drinking of the men, and the pipe-smoking of the women. The violent faction-fights which took place at the fairs which he frequented, were of a kind which he had never before observed among the pacific people of North Italy. These faction-fights were the result, partly of dram-drinking, and partly of the fighting mania which then prevailed in Ireland. There were also numbers of crippled and deformed beggars in every town,—quarrelling and fighting in the streets,—rows and drinkings at wakes,—gambling, duelling, and riotous living amongst all classes of the people,—things which could not but strike any ordinary observer at the time, but which have now, for the most part, happily passed away.

At the end of eighteen months, Bianconi's apprenticeship was out; and Faroni then offered to take him back to his father, in compliance with the original understanding. But Bianconi had no wish to return to Italy. Faroni then made over to him the money he had retained on his account, and Bianconi set up business for himself. He was now about eighteen years old; he was strong and healthy, and able to walk with a heavy load on his back from twenty to thirty miles a day. He bought a large case, filled it with coloured prints and other articles, and started from Dublin on a tour through the south of Ireland. He succeeded, like most persons who labour diligently. The curly-haired Italian lad became a general favourite. He took his native politeness with him everywhere; and made many friends among his various customers throughout the country.

Bianconi used to say that it was about this time when he was carrying his heavy case upon his back, weighing at least a hundred pounds—that the idea began to strike him, of some cheap method of conveyance being established for the accommodation of the poorer classes in Ireland. As he dismantled himself of his case of pictures, and sat wearied and resting on the milestones along the road, he puzzled his mind with the thought, "Why should poor people walk and toil, and rich people ride and take their ease? Could not some method be devised by which poor people also might have the opportunity of travelling comfortably?"

It will thus be seen that Bianconi was already beginning to think about the matter. When asked, not long before his death, how it was that he had first thought of starting his extensive Car establishment, he answered, "It grew out of my back!" It was the hundred weight of pictures on his dorsal muscles that stimulated his thinking faculties. But the time for starting his great experiment had not yet arrived.

Bianconi wandered about from town to town for nearly two years. The picture-case became heavier than ever. For a time he replaced it with a portfolio of unframed prints. Then he became tired of the wandering life, and in 1806 settled down at Carrick-on-Suir as a print-seller and carver and gilder. He supplied himself with gold-leaf from Waterford, to which town he used to proceed by Tom Morrissey's boat. Although the distance by road between the towns was only twelve miles, it was about twenty-four by water, in consequence of the windings of the river Suir. Besides, the boat could only go when the state of the tide permitted. Time was of little consequence; and it often took half a day to make the journey. In the course of one of his voyages, Bianconi got himself so thoroughly soaked by rain and mud that he caught a severe cold, which ran into pleurisy, and laid him up for about two months. He was carefully attended to by a good, kind physician, Dr. White, who would not take a penny for his medicine and nursing.

Business did not prove very prosperous at Carrick-on-suir; the town was small, and the trade was not very brisk. Accordingly, Bianconi resolved, after a year's ineffectual trial, to remove to Waterford, a more thriving centre of operations. He was now twenty-one years old. He began again as a carver and gilder; and as business flowed in upon him, he worked very hard, sometimes from six in the morning until two hours after midnight. As usual, he made many friends. Among the best of them was Edward Rice, the founder of the "Christian Brothers" in Ireland. Edward Rice was a true benefactor to his country. He devoted himself to the work of education, long before the National Schools were established; investing the whole of his means in the foundation and management of this noble institution.

Mr. Rice's advice and instruction set and kept Bianconi in the right road. He helped the young foreigner to learn English. Bianconi was no longer a dunce, as he had been at school; but a keen, active, enterprising fellow, eager to make his way in the world. Mr. Rice encouraged him to be sedulous and industrious, urged him to carefulness and sobriety, and strengthened his religions impressions. The help and friendship of this good man, operating upon the mind and soul of a young man, whose habits of conduct and whose moral and religious character were only in course of formation, could not fail to exercise, as Bianconi always acknowledged they did, a most powerful influence upon the whole of his after life.

Although "three removes" are said to be "as bad as a fire," Bianconi, after remaining about two years at Waterford, made a third removal in 1809, to Clonmel, in the county of Tipperary. Clonmel is the centre of a large corn trade, and is in water communication, by the Suir, with Carrick and Waterford. Bianconi, therefore, merely extended his connection; and still continued his dealings with his customers in the other towns. He made himself more proficient in the mechanical part of his business; and aimed at being the first carver and gilder in the trade. Besides, he had always an eye open for new business. At that time, when the war was raging with France, gold was at a premium. The guinea was worth about twenty-six or twenty-seven shillings. Bianconi therefore began to buy up the hoarded-up guineas of the peasantry. The loyalists became alarmed at his proceedings, and began to circulate the report that Bianconi, the foreigner, was buying up bullion to send secretly to Bonaparte! The country people, however, parted with their guineas readily; for they had no particular hatred of "Bony," but rather admired him.

Bianconi's conduct was of course quite loyal in the matter; he merely bought the guineas as a matter of business, and sold them at a profit to the bankers.

The country people had a difficulty in pronouncing his name. His shop was at the corner of Johnson Street, and instead of Bianconi, he came to be called "Bian of the Corner." He was afterwards known as "Bian."

Bianconi soon became well known after his business was established. He became a proficient in the carving and gilding line, and was looked upon as a thriving man. He began to employ assistants in his trade, and had three German gilders at work. While they were working in the shop he would travel about the country, taking orders and delivering goods—sometimes walking and sometimes driving.

He still retained a little of his old friskiness and spirit of mischief. He was once driving a car from Clonmel to Thurles; he had with him a large looking-glass with a gilt frame, on which about a fortnight's labour had been bestowed. In a fit of exuberant humour he began to tickle the horse under his tail with a straw! In an instant the animal reared and plunged, and then set off at a gallop down hill. The result was, that the car was dashed to bits and the looking-glass broken into a thousand atoms!

On another occasion, a man was carrying to Cashel on his back one of Bianconi's large looking-glasses. An old woman by the wayside, seeing the odd-looking, unwieldy package, asked what it was; on which Bianconi, who was close behind the man carrying the glass, answered that it was "the Repeal of the Union!" The old woman's delight was unbounded! She knelt down on her knees in the middle of the road, as if it had been a picture of the Madonna, and thanked God for having preserved her in her old age to see the Repeal of the Union!

But this little waywardness did not last long. Bianconi's wild oats were soon all sown. He was careful and frugal. As he afterwards used to say, "When I was earning a shilling a day at Clonmel, I lived upon eightpence." He even took lodgers, to relieve him of the charge of his household expenses. But as his means grew, he was soon able to have a conveyance of his own. He first started a yellow gig, in which he drove about from place to place, and was everywhere treated with kindness and hospitality. He was now regarded as "respectable," and as a person worthy to hold some local office. He was elected to a Society for visiting the Sick Poor, and became a Member of the House of Industry. He might have gone on in the same business, winning his way to the Mayoralty of Clonmel, which he afterwards held; but that the old idea, which had first sprung up in his mind while resting wearily on the milestones along the road, with his heavy case of pictures by his side, again laid hold of him, and he determined now to try whether his plan could not be carried into effect.

He had often lamented the fatigue that poor people had to undergo in travelling with burdens from place to place upon foot, and wondered whether some means might not be devised for alleviating their sufferings. Other people would have suggested "the Government!" Why should not the Government give us this, that, and the other,—give us roads, harbours, carriages, boats, nets, and so on. This, of course, would have been a mistaken idea; for where people are too much helped, they invariably lose the beneficent practice of helping themselves. Charles Bianconi had never been helped, except by advice and friendship. He had helped himself throughout; and now he would try to help others.

The facts were patent to everybody. There was not an Irishman who did not know the difficulty of getting from one town to another. There were roads between them, but no conveyances. There was an abundance of horses in the country, for at the close of the war an unusual number of horses, bred for the army, were thrown upon the market. Then a tax had been levied upon carriages, which sent a large number of jaunting-cars out of employment.

The roads of Ireland were on the whole good, being at that time quite equal, if not superior, to most of those in England. The facts of the abundant horses, the good roads, the number of unemployed outside cars, were generally known; but until Bianconi took the enterprise in hand, there was no person of thought, or spirit, or capital in the country, who put these three things together horses, roads, and cars and dreamt of remedying the great public inconvenience.

It was left for our young Italian carver and gilder, a struggling man of small capital, to take up the enterprise, and show what could be done by prudent action and persevering energy. Though the car system originally "grew out of his back," Bianconi had long been turning the subject over in his mind. His idea was, that we should never despise small interests, nor neglect the wants of poor people. He saw the mail-coaches supplying the requirements of the rich, and enabling them to travel rapidly from place to place. "Then," said he to himself, "would it not be possible for me to make an ordinary two-wheeled car pay, by running as regularly for the accommodation of poor districts and poor people?"

When Mr. Wallace, chairman of the Select Committee on Postage, in 1838, asked Mr. Bianconi, "What induced you to commence the car establishment?" his answer was, "I did so from what I saw, after coming to this country, of the necessity for such cars, inasmuch as there was no middle mode of conveyance, nothing to fill up the vacuum that existed between those who were obliged to walk and those who posted or rode. My want of knowledge of the language gave me plenty of time for deliberation, and in proportion as I grew up with the knowledge of the language and the localities, this vacuum pressed very heavily upon my mind, till at last I hit upon the idea of running jaunting-cars, and for that purpose I commenced running one between Clonmel and Cahir."[2]

What a happy thing it was for Bianconi and Ireland that he could not speak with facility,—that he did not know the language or the manners of the country! In his case silence was "golden." Had he been able to talk like the people about him, he might have said much and done little,—attempted nothing and consequently achieved nothing. He might have got up a meeting and petitioned Parliament to provide the cars, and subvention the car system; or he might have gone amongst his personal friends, asked them to help him, and failing their help, given up his idea in despair, and sat down grumbling at the people and the Government.

But instead of talking, he proceeded to doing, thereby illustrating Lessona's maxim of Volere e potere. After thinking the subject fully over, he trusted to self-help. He found that with his own means, carefully saved, he could make a beginning; and the beginning once made, included the successful ending.

The beginning, it is true, was very small. It was only an ordinary jaunting-car, drawn by a single horse, capable of accommodating six persons. The first car ran between Clonmel and Cahir, a distance of about twelve miles, on the 5th of July, 1815—a memorable day for Bianconi and Ireland. Up to that time the public accommodation for passengers was confined to a few mail and day coaches on the great lines of road, the fares by which were very high, and quite beyond the reach of the poorer or middle-class people.

People did not know what to make of Bianconi's car when it first started. There were, of course, the usual prophets of disaster, who decided that it "would never do." Many thought that no one would pay eighteen-pence for going to Cahir by car when they could walk there for nothing? There were others who thought that Bianconi should have stuck to his shop, as there was no connection whatever between picture-gilding and car-driving!

The truth is, the enterprise at first threatened to be a failure! Scarcely anybody would go by the car. People preferred trudging on foot, and saved their money, which was more valuable to them than their time. The car sometimes ran for weeks without a passenger. Another man would have given up the enterprise in despair. But this was not the way with Bianconi. He was a man of tenacity and perseverance. What should he do but start an opposition car? Nobody knew of it but himself; not even the driver of the opposition car. However, the rival car was started. The races between the car-drivers, the free lifts occasionally given to passengers, the cheapness of the fare, and the excitement of the contest, attracted the attention of the public. The people took sides, and before long both cars came in full. Fortunately the "great big yallah horse" of the opposition car broke down, and Bianconi had all the trade to himself.

The people became accustomed to travelling. They might still walk to Cahir; but going by car saved their legs, saved their brains, and saved their time. They might go to Cahir market, do their business there, and be comfortably back within the day. Bianconi then thought of extending the car to Tipperary and Limerick. In the course of the same year, 1815, he started another car between Clonmel, Cashel, and Thurles. Thus all the principal towns of Tipperary were, in the first year of the undertaking, connected together by car, besides being also connected with Limerick.

It was easy to understand the convenience of the car system to business men, farmers, and even peasants. Before their establishment, it took a man a whole day to walk from Thurles to Clonmel, the second day to do his business, and the third to walk back again; whereas he could, in one day, travel backwards and forwards between the two towns, and have five or six intermediate hours for the purpose of doing his business. Thus two clear days could be saved.

Still carrying out his scheme, Bianconi, in the following year (1816), put on a car from Clonmel to Waterford. Before that time there was no car accommodation between Clonmel and Carrick-on-Suir, about half-way to Waterford; but there was an accommodation by boat between Carrick and Waterford. The distance between the two latter places was, by road, twelve miles, and by the river Suir twenty-four miles. Tom Morrissey's boat plied two days a week; it carried from eight to ten passengers at 6 1/2d. of the then currency; it did the voyage in from four to five hours, and besides had to wait for the tide to float it up and down the river. When Bianconi's car was put on, it did the distance daily and regularly in two hours, at a fare of two shillings.

The people soon got accustomed to the convenience of the cars. They also learned from them the uses of punctuality and the value of time. They liked the open-air travelling and the sidelong motion. The new cars were also safe and well-appointed. They were drawn by good horses and driven by good coachmen. Jaunting-car travelling had before been rather unsafe. The country cars were of a ramshackle order, and the drivers were often reckless. "Will I pay the pike, or drive at it, plaise your honour?" said a driver to his passenger on approaching a turnpike-gate. Sam Lover used to tell a story of a car-driver, who, after driving his passenger up-hill and down-hill, along a very bad road, asked him for something extra at the end of his journey.

"Faith," said the driver, "its not putting me off with this ye'd be, if ye knew but all." The gentleman gave him another shilling. "And now what do you mean by saying, 'if ye knew but all?'" "That I druv yer honor the last three miles widout a linch-pin!"

Bianconi, to make sure of the soundness and safety of his cars, set up a workshop to build them for himself. He could thus depend upon their soundness, down even to the linch-pin itself. He kept on his carving and gilding shop until his car business had increased so much that it required the whole of his time and attention; and then he gave it up. In fact, when he was able to run a car from Clonmel to Waterford—a distance of thirty-two miles—at a fare of three-and-sixpence, his eventual triumph was secure.

He made Waterford one of the centres of his operations, as he had already made Clonmel. In 1818 he established a car between Waterford and Ross, in the following year a car between Waterford and Wexford, and another between Waterford and Enniscorthy. A few years later he established other cars between Waterford and Kilkenny, and Waterford and Dungarvan. From these furthest points, again, other cars were established in communication with them, carrying the line further north, east, and west. So much had the travelling between Clonmel and Waterford increased, that in a few years (instead of the eight or ten passengers conveyed by Tom Morrissey's boat on the Suir) there was horse and car power capable of conveying a hundred passengers daily between the two places.

Bianconi did a great stroke of business at the Waterford election of 1826. Indeed it was the turning point of his fortunes. He was at first greatly cramped for capital. The expense of maintaining and increasing his stock of cars, and of foddering his horses was very great; and he was always on the look-out for more capital. When the Waterford election took place, the Beresford party, then all-powerful, engaged all his cars to drive the electors to the poll. The popular party, however, started a candidate, and applied to Bianconi for help. But he could not comply, for his cars were all engaged. The morning after his refusal of the application, Bianconi was pelted with mud. One or two of his cars and horses were heaved over the bridge.

Bianconi then wrote to Beresford's agent, stating that he could no longer risk the lives of his drivers and his horses, and desiring to be released from his engagement. The Beresford party had no desire to endanger the lives of the car-drivers or their horses, and they set Bianconi free. He then engaged with the popular party, and enabled them to win the election. For this he was paid the sum of a thousand pounds. This access of capital was greatly helpful to him under the circumstances. He was able to command the market, both for horses and fodder. He was also placed in a position to extend the area of his car routes.

He now found time, amidst his numerous avocations, to get married! He was forty years of age before this event occurred. He married Eliza Hayes, some twenty years younger than himself, the daughter of Patrick Hayes, of Dublin, and of Henrietta Burton, an English-woman. The marriage was celebrated on the 14th of February, 1827; and the ceremony was performed by the late Archbishop Murray. Mr. Bianconi must now have been in good circumstances, as he settled two thousand pounds upon his wife on their marriage-day. His early married life was divided between his cars, electioneering, and Repeal agitation—for he was always a great ally of O'Connell. Though he joined in the Repeal movement, his sympathies were not with it; for he preferred Imperial to Home Rule. But he could never deny himself the pleasure of following O'Connell, "right or wrong."

Let us give a picture of Bianconi now. The curly-haired Italian boy had grown a handsome man. His black locks curled all over his head like those of an ancient Roman bust. His face was full of power, his chin was firm, his nose was finely cut and well-formed; his eyes were keen and sparkling, as if throwing out a challenge to fortune. He was active, energetic, healthy, and strong, spending his time mostly in the open air. He had a wonderful recollection of faces, and rarely forgot to recognise the countenance that he had once seen. He even knew all his horses by name. He spent little of his time at home, but was constantly rushing about the country after business, extending his connections, organizing his staff, and arranging the centres of his traffic.

To return to the car arrangements. A line was early opened from Clonmel—which was at first the centre of the entire connection—to Cork; and that line was extended northward, through Mallow and Limerick. Then, the Limerick car went on to Tralee, and from thence to Cahirciveen, on the south-west coast of Ireland. The cars were also extended northward from Thurles to Roscrea, Ballinasloe, Athlone, Roscommon, and Sligo, and to all the principal towns in the north-west counties of Ireland.

The cars interlaced with each other, and plied, not so much in continuous main lines, as across country, so as to bring all important towns, but especially the market towns, into regular daily communication with each other. Thus, in the course of about thirty years, Bianconi succeeded in establishing a system of internal communication in Ireland, which traversed the main highways and cross-roads from town to town, and gave the public a regular and safe car accommodation at the average rate of a penny-farthing per mile.

The traffic in all directions steadily increased. The first car used was capable of accommodating only six persons. This was between Clonmel and Cahir. But when it went on to Limerick, a larger car was required. The traffic between Clonmel and Waterford was also begun with a small-sized car. But in the course of a few years, there were four large-sized cars, travelling daily each way, between the two places. And so it was in other directions, between Cork in the south; and Sligo and Strabane in the north and north-west; between Wexford in the east, and Galway and Skibbereen in the west and south-west.

Bianconi first increased the accommodation of these cars so as to carry four persons on each side instead of three, drawn by two horses. But as the two horses could quite as easily carry two additional passengers, another piece was added to the car so as to carry five passengers. Then another four-wheeled car was built, drawn by three horses, so as to carry six passengers on each side. And lastly, a fourth horse was used, and the car was further enlarged, so as to accommodate seven, and eventually eight passengers on each side, with one on the box, which made a total accommodation for seventeen passengers. The largest and heaviest of the long cars, on four wheels, was called "Finn MacCoul's," after Ossian's Giant; the fast cars, of a light build, on two wheels, were called "Faugh-a-ballagh," or "clear the way"; while the intermediate cars were named "Massey Dawsons," after a popular Tory squire.

When Bianconi's system was complete, he had about a hundred vehicles at work; a hundred and forty stations for changing horses, where from one to eight grooms were employed; about a hundred drivers, thirteen hundred horses, performing an average distance of three thousand eight hundred miles daily; passing through twenty-three counties, and visiting no fewer than a hundred and twenty of the principal towns and cities in the south and west and midland counties of Ireland. Bianconi's horses consumed on an average from three to four thousand tons of hay yearly, and from thirty to forty thousand barrels of oats, all of which were purchased in the respective localities in which they were grown.

Bianconi's cars—or "The Bians"—soon became very popular. Everybody was under obligations to them. They greatly promoted the improvement of the country. People could go to market and buy or sell their goods more advantageously. It was cheaper for them to ride than to walk. They brought the whole people of the country so much nearer to each other. They virtually opened up about seven-tenths of Ireland to civilisation and commerce, and among their other advantages, they opened markets for the fresh fish caught by the fishermen of Galway, Clifden, Westport, and other places, enabling them to be sold throughout the country on the day after they were caught. They also opened the magnificent scenery of Ireland to tourists, and enabled them to visit Bantry Bay, Killarney, South Donegal, and the wilds of Connemara in safety, all the year round.

Bianconi's service to the public was so great, and it was done with so much tact, that nobody had a word to say against him. Everybody was his friend. Not even the Whiteboys would injure him or the mails he carried. He could say with pride, that in the most disturbed times his cars had never been molested. Even during the Whiteboy insurrection, though hundreds of people were on the roads at night, the traffic went on without interference. At the meeting of the British Association in 1857, Bianconi said: "My conveyances, many of them carrying very important mails, have been travelling during all hours of the day and night, often in lonely and unfrequented places; and during the long period of forty-two years that my establishment has been in existence, the slightest injury has never been done by the people to my property, or that entrusted to my care; and this fact gives me greater pleasure than any pride I might feel in reflecting upon the other rewards of my life's labour."

Of course Bianconi's cars were found of great use for carrying the mails. The post was, at the beginning of his enterprise, very badly served in Ireland, chiefly by foot and horse posts. When the first car was run from Clonmel to Cahir, Bianconi offered to carry the mail for half the price then paid for "sending it alternately by a mule and a bad horse." The post was afterwards found to come regularly instead of irregularly to Cahir; and the practice of sending the mails by Bianconi's cars increased from year to year. Dispatch won its way to popularity in Ireland as elsewhere, and Bianconi lived to see all the cross-posts in Ireland arranged on his system.

The postage authorities frequently used the cars of Bianconi as a means of competing with the few existing mail-coaches. For instance, they asked him to compete for carrying the post between Limerick and Tralee, then carried by a mail-coach. Before tendering, Bianconi called on the contractor, to induce him to give in to the requirements of the Post Office, because he knew that the postal authorities only desired to make use of him to fight the coach proprietors. But having been informed that it was the intention of the Post Office to discontinue the mail-coach whether Bianconi took the contract or not, he at length sent in his tender, and obtained the contract.

He succeeded in performing the service, and delivered the mail much earlier than it had been done before. But the former contractor, finding that he had made a mistake, got up a movement in favour of re-establishing the mail-coach upon that line of road; and he eventually induced the postage authorities to take the mail contract out of the hands of Bianconi, and give it back to himself, as formerly. Bianconi, however, continued to keep his cars upon the road. He had before stated to the contractor, that if he once started his cars, he would not leave it, even though the contract were taken from him. Both coach and car therefore ran for years upon the road, each losing thousands of pounds. "But," said Bianconi, when asked about the matter by the Committee on Postage in 1838, "I kept my word: I must either lose character by breaking my word, or lose money. I prefer losing money to giving up the line of road."

Bianconi had also other competitors to contend with, especially from coach and car proprietors. No sooner had he shown to others the way to fortune, than he had plenty of imitators. But they did not possess his rare genius for organisation, nor perhaps his still rarer principles. They had not his tact, his foresight, his knowledge, nor his perseverance. When Bianconi was asked by the Select Committee on Postage, "Do the opposition cars started against you induce you to reduce your fares?" his answer was, "No; I seldom do. Our fares are so close to the first cost, that if any man runs cheaper than I do, he must starve off, as few can serve the public lower and better than I do."[3]

Bianconi was once present at a meeting of car proprietors, called for the purpose of uniting to put down a new opposition coach. Bianconi would not concur, but protested against it, saying, "If car proprietors had united against me when I started, I should have been crushed. But is not the country big enough for us all?" The coach proprietors, after many angry words, threatened to unite in running down Bianconi himself. "Very well," he said, "you may run me off the road—that is possible; but while there is this" (pulling a flower out of his coat) "you will not put me down." The threat merely ended in smoke, the courage and perseverance of Bianconi having long since become generally recognised.

We have spoken of the principles of Mr. Bianconi. They were most honourable. His establishment might be spoken of as a school of morality. In the first place, he practically taught and enforced the virtues of punctuality, truthfulness, sobriety, and honesty. He also taught the public generally the value of time, to which, in fact, his own success was in a great measure due. While passing through Clonmel in 1840, Mr. and Mrs. S. C. Hall called upon Bianconi and went over his establishment, as well as over his house and farm, a short distance from the town. The travellers had a very pressing engagement, and could not stay to hear the story of how their entertainer had contrived to "make so much out of so little." "How much time have you?" he asked. "Just five minutes." "The car," says Mr. Hall, "had conveyed us to the back entrance. Bianconi instantly rang the bell, and said to the servant, 'Tell the driver to bring the car round to the front,' adding, 'that will save one minute, and enable me to tell you all within the time.' This was, in truth the secret of his success, making the most of time."[4]

But the success of Bianconi was also due to the admirable principles on which his establishment was conducted. His drivers were noted as being among the most civil and obliging men in Ireland, besides being pleasant companions to boot. They were careful, punctual, truthful, and honest; but all this was the result of strict discipline on the part of their master.

The drivers were taken from the lowest grades of the establishment, and promoted to higher positions according to their respective merits as opportunity offered. "Much surprise," says Bianconi, "has often been expressed at the high order of men connected with my car establishment and at its popularity; but parties thus expressing themselves forget to look at Irish society with sufficient grasp. For my part, I cannot better compare it than to a man merging to convalescence from a serious attack of malignant fever, and requiring generous nutrition in place of medical treatment"[5]

To attach the men to the system, as well as to confer upon them the due reward for their labour, he provided for all the workmen who had been injured, worn out, or become superannuated in his service. The drivers could then retire upon a full pension, which they enjoyed during the rest of their lives. They were also paid their full wages during sickness, and at their death Bianconi educated their children, who grew up to manhood, and afterwards filled the situations held by their deceased parents.

Every workman had thus a special interest in his own good conduct. They knew that nothing but misbehaviour could deprive them of the benefits they enjoyed; and hence their endeavours to maintain their positions by observing the strict discipline enjoined by their employer.

Sobriety was, of course, indispensable—a drunken car-driver being amongst the most dangerous of servants. The drivers must also be truthful, and the man found telling a lie, however venial, was instantly dismissed. Honesty was also strongly enforced, not only for the sake of the public, but for the sake of the men themselves. Hence he never allowed his men to carry letters. If they did so, he fined them in the first instance very severely, and in the second instance dismissed them. "I do so," he said, "because if I do not respect other institutions (the Post Office), my men will soon learn not to respect my own. Then, for carrying letters during the extent of their trip, the men most probably would not get money, but drink, and hence become dissipated and unworthy of confidence."

Thus truth, accuracy, punctuality, sobriety, and honesty being strictly enforced, formed the fundamental principle of the entire management. At the same time, Bianconi treated his drivers with every confidence and respect. He made them feel that, in doing their work well, they conferred a greater benefit on him and on the public than he did on them by paying them their wages.

When attending the British Association at Cork, Bianconi said that, "in proportion as he advanced his drivers, he lowered their wages." "Then," said Dr. Taylor, the Secretary, "I wouldn't like to serve you." "Yes, you would," replied Bianconi, "because in promoting my drivers I place them on a more lucrative line, where their certainty of receiving fees from passengers is greater."

Bianconi was as merciful to his horses as to his men. He had much greater difficulty at first in finding good men than good horses, because the latter were not exposed to the temptations to which the former were subject. Although the price of horses continued to rise, he nevertheless bought the best horses at increased prices, and he took care not to work them overmuch. He gave his horses as well as his men their seventh day's rest. "I find by experience," he said, "that I can work a horse eight miles a day for six days in the week, easier than I can work six miles for seven days; and that is one of my reasons for having no cars, unless carrying a mail, plying upon Sundays."

Bianconi had confidence in men generally. The result was that men had confidence in him. Even the Whiteboys respected him. At the close of a long and useful life he could say with truth, "I never yet attempted to do an act of generosity or common justice, publicly or privately, that I was not met by manifold reciprocity."

By bringing the various classes of society into connection with each other, Bianconi believed, and doubtless with truth, that he was the means of making them respect each other, and that he thereby promoted the civilisation of Ireland. At the meeting of the social Science Congress, held at Dublin in 1861, he said: "The state of the roads was such as to limit the rate of travelling to about seven miles an hour, and the passengers were often obliged to walk up hills. Thus all classes were brought together, and I have felt much pleasure in believing that the intercourse thus created tended to inspire the higher classes with respect and regard for the natural good qualities of the humbler people, which the latter reciprocated by a becoming deference and an anxiety to please and oblige. Such a moral benefit appears to me to be worthy of special notice and congratulation."

Even when railways were introduced, Bianconi did not resist them, but welcomed them as "the great civilisers of the age." There was, in his opinion, room enough for all methods of conveyance in Ireland. When Captain Thomas Drummond was appointed Under-Secretary for Ireland in 1835, and afterwards chairman of the Irish Railway Commission, he had often occasion to confer with Mr. Bianconi, who gave him every assistance. Mr. Drummond conceived the greatest respect for Bianconi, and often asked him how it was that he, a foreigner, should have acquired so extensive an influence and so distinguished a position in Ireland?

"The question came upon me," said Bianconi, "by surprise, and I did not at the time answer it. But another day he repeated his question, and I replied, 'Well, it was because, while the big and the little were fighting, I crept up between them, carried out my enterprise, and obliged everybody.'" This, however, did not satisfy Mr. Drummond, who asked Bianconi to write down for him an autobiography, containing the incidents of his early life down to the period of his great Irish enterprise. Bianconi proceeded to do this, writing down his past history in the occasional intervals which he could snatch from the immense business which he still continued personally to superintend. But before the "Drummond memoir" could be finished Mr. Drummond himself had ceased to live, having died in 1840, principally of overwork. What he thought of Bianconi, however, has been preserved in his Report of the Irish Railway Commission of 1838, written by Mr. Drummond himself, in which he thus speaks of his enterprising friend in starting and conducting the great Irish car establishment:—

"With a capital little exceeding the expense of outfit he commenced. Fortune, or rather the due reward of industry and integrity, favoured his first efforts. He soon began to increase the number of his cars and multiply routes, until his establishment spread over the whole of Ireland. These results are the more striking and instructive as having been accomplished in a district which has long been represented as the focus of unreclaimed violence and barbarism, where neither life nor property can be deemed secure. Whilst many possessing a personal interest in everything tending to improve or enrich the country have been so misled or inconsiderate as to repel by exaggerated statements British capital from their doors, this foreigner chose Tipperary as the centre of his operations, wherein to embark all the fruits of his industry in a traffic peculiarly exposed to the power and even to the caprice of the peasantry. The event has shown that his confidence in their good sense was not ill-grounded.

"By a system of steady and just treatment he has obtained a complete mastery, exempt from lawless intimidation or control, over the various servants and agents employed by him, and his establishment is popular with all classes on account of its general usefulness and the fair liberal spirit of its management. The success achieved by this spirited gentleman is the result, not of a single speculation, which might have been favoured by local circumstances, but of a series of distinct experiments, all of which have been successful."

When the railways were actually made and opened, they ran right through the centre of Bianconi's long-established systems of communication. They broke up his lines, and sent them to the right and left. But, though they greatly disturbed him, they did not destroy him. In his enterprising hands the railways merely changed the direction of the cars. He had at first to take about a thousand horses off the road, with thirty-seven vehicles, travelling 2446 miles daily. But he remodelled his system so as to run his cars between the railway-stations and the towns to the right and left of the main lines.

He also directed his attention to those parts of Ireland which had not before had the benefit of his conveyances. And in thus still continuing to accommodate the public, the number of his horses and carriages again increased, until, in 1861, he was employing 900 horses, travelling over 4000 miles daily; and in 1866, when he resigned his business, he was running only 684 miles daily below the maximum run in 1845, before the railways had begun to interfere with his traffic.

His cars were then running to Dungarvan, Waterford, and Wexford in the south-west of Ireland; to Bandon, Rosscarbery, Skibbereen, and Cahirciveen, in the south; to Tralee, Galway, Clifden, Westport, and Belmullet in the west; to Sligo, Enniskillen, Strabane, and Letterkenny in the north; while, in the centre of Ireland, the towns of Thurles, Kilkenny, Birr, and Ballinasloe were also daily served by the cars of Bianconi.

At the meeting of the British Association, held in Dublin in 1857, Mr. Bianconi mentioned a fact which, he thought, illustrated the increasing prosperity of the country and the progress of the people. It was, that although the population had so considerably decreased by emigration and other causes, the proportion of travellers by his conveyances continued to increase, demonstrating not only that the people had more money, but that they appreciated the money value of time, and also the advantages of the car system established for their accommodation.

Although railways must necessarily have done much to promote the prosperity of Ireland, it is very doubtful whether the general passenger public were not better served by the cars of Bianconi than by the railways which superseded them. Bianconi's cars were on the whole cheaper, and were always run en correspondence, so as to meet each other; whereas many of the railway trains in the south of Ireland, under the competitive system existing between the several companies, are often run so as to miss each other. The present working of the Irish railway traffic provokes perpetual irritation amongst the Irish people, and sufficiently accounts for the frequent petitions presented to Parliament that they should be taken in hand and worked by the State.

Bianconi continued to superintend his great car establishment until within the last few years. He had a constitution of iron, which he expended in active daily work. He liked to have a dozen irons in the fire, all red-hot at once. At the age of seventy he was still a man in his prime; and he might be seen at Clonmel helping, at busy times, to load the cars, unpacking and unstrapping the luggage where it seemed to be inconveniently placed; for he was a man who could never stand by and see others working without having a hand in it himself. Even when well on to eighty, he still continued to grapple with the immense business involved in working a traffic extending over two thousand five hundred miles of road.

Nor was Bianconi without honour in his adopted country. He began his great enterprise in 1815, though it was not until 1831 that he obtained letters of naturalisation. His application for these privileges was supported by the magistrates of Tipperary and by the Grand Jury, and they were at once granted. In 1844 he was elected Mayor of Clonmel, and took his seat as Chairman at the Borough Petty Sessions to dispense justice.

The first person brought before him was James Ryan, who had been drunk and torn a constable's belt. "Well, Ryan," said the magistrate, "what have you to say?" "Nothing, your worship; only I wasn't drunk." "Who tore the constable's belt?" "He was bloated after his Christmas dinner, your worship, and the belt burst!" "You are so very pleasant," said the magistrate, "that you will have to spend forty-eight hours in gaol."

He was re-elected Mayor in the following year, very much against his wish. He now began to buy land, for "land hunger" was strong upon him. In 1846 he bought the estate of Longfield, in the parish of Boherlahan, county of Tipperary. It consisted of about a thousand acres of good land, with a large cheerful house overlooking the river Suir. He went on buying more land, until he became possessor of about eight thousand English acres.

One of his favourite sayings was: "Money melts, but land holds while grass grows and water runs." He was an excellent landlord, built comfortable houses for his tenantry, and did what he could for their improvement. Without solicitation, the Government appointed him a justice of the peace and a Deputy-lientenant for the county of Tipperary. Everything that he did seemed to thrive. He was honest, straightforward, loyal, and law-abiding.

On first taking possession of his estate at Longfield, he was met by a procession of the tenantry, who received him with great enthusiasm. In his address to them, he said, amongst other things: "Allow me to impress upon you the great importance of respecting the laws. The laws are made for the good and the benefit of society, and for the punishment of the wicked. No one but an enemy would counsel you to outrage the laws. Above all things, avoid secret and unlawful societies. Much of the improvement now going on amongst us is owing to the temperate habits of the people, to the mission of my much respected friend, Father Mathew, and to the advice of the Liberator. Follow the advice of O'Connell; be temperate, moral, peaceable; and you will advance your country, ameliorate your condition, and the blessing of God will attend all your efforts."

Bianconi was always a great friend of O'Connell. From an early period he joined him in the Catholic Emancipation movement. He took part with him in founding the National Bank in Ireland. In course of time the two became more intimately related. Bianconi's son married O'Connell's granddaughter; and O'Connell's nephew, Morgan John, married Bianconi's daughter. Bianconi's son died in 1864, leaving three daughters, but no male heir to carry on the family name. The old man bore the blow of his son's premature death with fortitude, and laid his remains in the mortuary chapel, which he built on his estate at Longfield.

In the following year, when he was seventy-eight, he met with a severe accident. He was overturned, and his thigh was severely fractured. He was laid up for six months, quite incapable of stirring. He was afterwards able to get about in a marvellous way, though quite crippled. As his life's work was over, he determined to retire finally from business; and he handed over the whole of his cars, coaches, horses, and plant, with all the lines of road he was then working, to his employes, on the most liberal terms.

My youngest son met Mr. Bianconi, by appointment, at the Roman Catholic church at Boherlahan, in the summer of 1872. Although the old gentleman had to be lifted into and out of his carriage by his two men-servants, he was still as active-minded as ever. Close to the church at Boherlahan is Bianconi's mortuary chapel, which he built as a sort of hobby, for the last resting-place of himself and his family. The first person interred in it was his eldest daughter, who died in Italy; the second was his only son. A beautiful monument with a bas-relief has been erected in the chapel by Benzoni, an Italian sculptor, to the memory of his daughter.

"As we were leaving the chapel," my son informs me, "we passed a long Irish car containing about sixteen people, the tenants of Mr. Bianconi, who are brought at his expense from all parts of the estate. He is very popular with his tenantry, regarding their interests as his own; and he often quotes the words of his friend Mr. Drummond, that 'property has its duties as well as its rights.' He has rebuilt nearly every house on his extensive estates in Tipperary.

"On our way home, the carriage stopped to let me down and see the strange remains of an ancient fort, close by the roadside. It consists of a high grass-grown mound, surrounded by a moat. It is one of the so-called Danish forts, which are found in all parts of Ireland. If it be true that these forts were erected by the Danes, they must at one time have had a strong hold of the greater part of Ireland.

"The carriage entered a noble avenue of trees, with views of prettily enclosed gardens on either side. Mr. Bianconi exclaimed, 'Welcome to the Carman's Stage!' Longfield House, which we approached, is a fine old-fashioned house, situated on the river Suir, a few miles south of Cashel, one of the most ancient cities in Ireland. Mr. Bianconi and his family were most hospitable; and I found him most lively and communicative. He talked cleverly and with excellent choice of language for about three hours, during which I learnt much from him.

"Like most men who have accomplished great things, and overcome many difficulties, Mr. Bianconi is fond of referring to the past events in his interesting life. The acuteness of his conversation is wonderful. He hits off a keen thought in a few words, sometimes full of wit and humour. I thought this very good: 'Keep before the wheels, young man, or they will run over you: always keep before the wheels!' He read over to me the memoir he had prepared at the suggestion of Mr. Drummond, relating to the events of his early life; and this opened the way for a great many other recollections not set down in the book.

"He vividly remembered the parting from his mother, nearly seventy years ago, and spoke of her last words to him: 'When you remember me, think of me as waiting at this window, watching for your return.' This led him to speak of the great forgetfulness and want of respect which children have for their parents nowadays. 'We seem,' he said, 'to have fallen upon a disrespectful age.'

"'It is strange,' said he, 'how little things influence one's mind and character. When I was a boy at Waterford, I bought an old second-hand book from a man on the quay, and the maxim on its title-page fixed itself deeply on my memory. It was, "Truth, like water, will find its own level."' And this led him to speak of the great influence which the example and instruction of Mr. Rice, of the Christian Brothers, had had upon his mind and character. 'That religions institution,' said he, 'of which Mr. Rice was one of the founders, has now spread itself over the country, and, by means of the instruction which the members have imparted to the poorer ignorant classes, they have effected quite a revolution in the south of Ireland.'

"'I am not much of a reader,' said Mr. Bianconi; 'the best part of my reading has consisted in reading way-bills. But I was once complimented by Justice Lefroy upon my books. He remarked to me what a wonderful education I must have had to invent my own system of book-keeping. Yes,' said he, pointing to his ledgers, 'there they are.' The books are still preserved, recording the progress of the great car enterprise. They show at first the small beginnings, and then the rapid growth—the tens growing to hundreds, and the hundreds to thousands—the ledgers and day-books containing, as it were, the whole history of the undertaking—of each car, of each man, of each horse, and of each line of road, recorded most minutely.

"'The secret of my success,' said he, 'has been promptitude, fair dealing, and good humour. And this I will add, what I have often said before, that I never did a kind action but it was returned to me tenfold. My cars have never received the slightest injury from the people. Though travelling through the country for about sixty years, the people have throughout respected the property intrusted to me. My cars have passed through lonely and unfrequented places, and they have never, even in the most disturbed times, been attacked. That, I think, is an extraordinary testimony to the high moral character of the Irish people.'

"'It is not money, but the genius of money that I esteem,' said Bianconi; 'not money itself, but money used as a creative power.'

And he himself has furnished in his own life the best possible illustration of his maxim He created a new industry, gave employment to an immense number of persons, promoted commerce, extended civilisation; and, though a foreigner, proved one of the greatest of Ireland's benefactors."

About two years after the date of my son's visit, Charles Bianconi passed away, full of years and honours; and his remains were laid beside those of his son and daughter, in the mortuary chapel at Boherlahan. He died in 1875, in his ninetieth year. Well might Signor Henrico Mayer say, at the British Association at Cork in 1846, that "he felt proud as an Italian to hear a compatriot so deservedly eulogised; and although Ireland might claim Bianconi as a citizen, yet the Italians should ever with pride hail him as a countryman, whose industry and virtue reflected honour on the country of his birth."

Footnotes for Chapter IX.

[1] This article originally appeared in 'Good Words.' A biography of Charles Bianconi, by his daughter, Mrs. Morgan John O'Connell, has since been published; but the above article is thought worthy of republication, as its contents were for the most part taken principally from Mr. Bianconi's own lips.

[2] Minutes of Evidence taken before the Select Committee on Postage (Second Report), 1838, p. 284.

[3] Evidence before the Select Committee on Postage, 1838.

[4] Hall's 'Ireland,' ii. 76.

[5] Paper read before the British Association at Cork, 1843.

CHAPTER X.

INDUSTRY IN IRELAND: THROUGH CONNAUGHT AND ULSTER, TO BELFAST.

"The Irish people have a past to boast of, and a future to create."—J. F. O'Carrol.

"One of the great questions is how to find an outlet for Irish manufactures. We ought to be an exporting nation, or we never will be able to compete successfully with our trade rivals."—E. D. Gray.

"Ireland may become a Nation again, if we all sacrifice our parricidal passions, prejudices, and resentments on the altar of our country. Then shall your manufactures flourish, and Ireland be free."—Daniel O'Connell.

Further communications passed between my young friend, the Italian count, and his father; and the result was that he accompanied me to Ireland, on the express understanding that he was to send home a letter daily by post assuring his friends of his safety. We went together accordingly to Galway, up Lough Corrib to Cong and Lough Mask; by the romantic lakes and mountains of Connemara to Clifden and Letterfrack, and through the lovely pass of Kylemoor to Leenane; along the fiord of Killury; then on, by Westport and Ballina to Sligo. Letters were posted daily by my young friend; and every day we went forwards in safety.

But how lonely was the country! We did not meet a single American tourist during the whole course of our visit, and the Americans are the most travelling people in the world. Although the railway companies have given every facility for visiting Connemara and the scenery of the West of Ireland, we only met one single English tourist, accompanied by his daughter. The Bianconi long car between Clifden and Westport had been taken off for want of support. The only persons who seemed to have no fear of Irish agrarianism were the English anglers, who are ready to brave all dangers, imaginary or supposed, provided they can only kill a big salmon! And all the rivers flowing westward into the Atlantic are full of fine fish. While at Galway, we looked down into the river Corrib from the Upper Bridge, and beheld it literally black with the backs of salmon! They were waiting for a flood to enable them to ascend the ladder into Lough Corrib. While there, 1900 salmon were taken in one day by nets in the bay.

Galway is a declining town. It has docks, but no shipping; bonded warehouses, but no commerce. It has a community of fishermen at Claddagh, but the fisheries of the bay are neglected. As one of the poor men of the place exclaimed, "Poverty is the curse of Ireland." On looking at Galway from the Claddagh side, it seems as if to have suffered from a bombardment. Where a roof has fallen in, nothing has been done to repair it. It was of no use. The ruin has been left to go on. The mills, which used to grind home-grown corn, are now unemployed. The corn comes ready ground from America. Nothing is thought of but emigration, and the best people are going, leaving the old, the weak, and the inefficient at home. "The labourer," said the late President Garfield, "has but one commodity to sell—his day's work, it is his sole reliance. He must sell it to-day, or it is lost for-ever." And as the poor Irishman cannot sell his day's labour, he must needs emigrate to some other country, where his only commodity may be in demand.

While at Galway, I read with interest an eloquent speech delivered by Mr. Parnell at the banquet held in the Great Hall of the Exhibition at Cork. Mr. Parnell asked, with much reason, why manufactures should not be established and encouraged in the South of Ireland, as in other parts of the country. Why should not capital be invested, and factories and workshops developed, through the length and breadth of the kingdom? "I confess," he said, "I should like to give Ireland a fair opportunity of working her home manufactures. We can each one of us do much to revive the ancient name of our nation in those industrial pursuits which have done so much to increase and render glorious those greater nations by the side of which we live. I trust that before many years are over we shall have the honour and pleasure of meeting in even a more splendid palace than this, and of seeing in the interval that the quick-witted genius of the Irish race has profited by the lessons which this beautiful Exhibition must undoubtedly teach, and that much will have been done to make our nation happy, prosperous, and free."

Mr. Parnell, in the course of his speech, referred to the manufactures which had at one time flourished in Ireland—to the flannels of Rathdrum, the linens of Bandon, the cottons of Cork, and the gloves of Limerick. Why should not these things exist again? "We have a people who are by nature quick and facile to learn, who have shown in many other countries that they are industrious and laborious, and who have not been excelled—whether in the pursuits of agriculture under a midday sun in the field, or amongst the vast looms in the factory districts—by the people of any country on the face of the globe."[1] Most just and eloquent!

The only weak point in Mr. Parnell's speech was where he urged his audience "not to use any article of the manufacture of any other country except Ireland, where you can get up an Irish manufacture." The true remedy is to make Irish articles of the best and cheapest, and they will be bought, not only by the Irish, but by the English and people of all nations. Manufactures cannot be "boycotted." They will find their way into all lands, in spite even of the most restrictive tariffs. Take, for instance, the case of Belfast hereafter to be referred to. If the manufacturing population of that town were to rely for their maintenance on the demand for their productions at home, they would simply starve. But they make the best and the cheapest goods of their kind, and hence the demand for them is world-wide.

There is an abundant scope for the employment of capital and skilled labour in Ireland. During the last few years land has been falling rapidly out of cultivation. The area under cereal crops has accordingly considerably decreased.[2] Since 1868, not less than 400,000 acres have been disused for this purpose.[3] Wheat can be bought better and cheaper in America, and imported into Ireland ground into flour. The consequence is, that the men who worked the soil, as well as the men who ground the corn, are thrown out of employment, and there is nothing left for them but subsistence upon the poor-rates, emigration to other countries, or employment in some new domestic industry.

Ireland is by no means the "poor Ireland" that she is commonly supposed to be. The last returns of the Postmaster-General show that she is growing in wealth. Irish thrift has been steadily at work during the last twenty years. Since the establishment of the Post Office Savings Banks, in 1861, the deposits have annually increased in value. At the end of 1882, more than two millions sterling had been deposited in these banks, and every county participated in the increase.[4] The largest accumulations were in the counties of Dublin, Antrim, Cork, Down, Tipperary, and Tyrone, in the order named. Besides this amount, the sum of 2,082,413L. was due to depositors in the ordinary Savings Banks on the 20th of November, 1882; or, in all, more than four millions sterling, the deposits of small capitalists. At Cork, at the end of last year, it was found that the total deposits made in the savings bank had been 76,000L, or an increase of 6,675L. over the preceding twelve months. But this is not all. The Irish middle classes are accustomed to deposit most of their savings in the Joint Stock banks; and from the returns presented to the Lord Lieutenant, dated the 31st of January, 1883, we find that these had been more than doubled in twenty years, the deposits and cash balances having increased from 14,389,000L. at the end of 1862, to 32,746,000L. at the end of 1882. During the last year they had increased by the sum of 2,585,000L. "So large an increase in bank deposits and cash balances," says the Report, "is highly satisfactory." It may be added that the investments in Government and India Stock, on which dividends were paid at the Bank of Ireland, at the end of 1882, amounted to not less than 31,804,000L.

It is proper that Ireland should be bountiful with her increasing means. It has been stated that during the last eighteen years her people have contributed not less than six millions sterling for the purpose of building places of worship, convents, schools, and colleges, in connection with the Roman Catholic Church, not to speak of their contributions for other patriotic objects.

It would be equally proper if some of the saved surplus capital of Ireland, as suggested by Mr. Parnell, were invested in the establishment of Irish manufactures. This would not only give profitable occupation to the unemployed, but enable Ireland to become an increasingly exporting nation. We are informed by an Irish banker, that there is abundance of money to be got in Ireland for any industry which has a reasonable chance of success. One thing, however, is certain: there must be perfect safety. An old writer has said that "Government is a badge of lost innocence: the palaces of kings are built upon the ruins of the bowers of paradise." The main use of government is protection against the weaknesses and selfishness of human nature. If there be no protection for life, liberty, property, and the fruits of accumulated industry, government becomes comparatively useless, and society is driven back upon its first principles.

Capital is the most sensitive of all things. It flies turbulence and strife, and thrives only in security and freedom. It must have complete safety. If tampered with by restrictive laws, or hampered by combinations, it suddenly disappears. "The age of glory of a nation," said Sir Humphry Davy, "is the age of its security. The same dignified feeling which urges men to gain a dominion over nature will preserve them from the dominion of slavery. Natural, and moral, and religions knowledge, are of one family; and happy is the country and great its strength where they dwell together in union."

Dublin was once celebrated for its shipbuilding, its timber-trade, its iron manufactures, and its steam-printing; Limerick was celebrated for its gloves; Kilkenny for its blankets; Bandon for its woollen and linen manufactures. But most of these trades were banished by strikes.[5] Dr. Doyle stated before the Irish Committee of 1830, that the almost total extinction of the Kilkenny blanket-trade was attributable to the combinations of the weavers; and O'Connell admitted that Trades Unions had wrought more evil to Ireland than absenteeism and Saxon maladministration. But working men have recently become more prudent and thrifty; and it is believed that under the improved system of moderate counsel, and arbitration between employers and employed, a more hopeful issue is likely to attend the future of such enterprises.

Another thing is clear. A country may be levelled down by idleness and ignorance; it can only be levelled up by industry and intelligence. It is easy to pull down; it is very difficult to build up. The hands that cannot erect a hovel may demolish a palace. We have but to look to Switzerland to see what a country may become which mixes its industry with its brains. That little land has no coal, no seaboard by which she can introduce it, and is shut off from other countries by lofty mountains, as well as by hostile tariffs; and yet Switzerland is one of the most prosperous nations in Europe, because governed and regulated by intelligent industry. Let Ireland look to Switzerland, and she need not despair.

Ireland is a much richer country by nature than is generally supposed. In fact, she has not yet been properly explored. There is copper-ore in Wicklow, Waterford, and Cork. The Leitrim iron-ores are famous for their riches; and there is good ironstone in Kilkenny, as well as in Ulster. The Connaught ores are mixed with coal-beds. Kaolin, porcelain clay, and coarser clay, abound; but it is only at Belleek that it has been employed in the pottery manufacture. But the sea about Ireland is still less explored than the land. All round the Atlantic seaboard of the Irish coast are shoals of herring and mackerel, which might be food for men, but are at present only consumed by the multitudes of sea-birds which follow them.

In the daily papers giving an account of the Cork Exhibition, appeared the following paragraph: "An interesting exhibit will be a quantity of preserved herrings from Lowestoft, caught off the old head of Kinsale, and returned to Cork after undergoing a preserving process in England."[6] Fish caught off the coast of Ireland by English fishermen, taken to England and cured, and then "returned to Cork" for exhibition! Here is an opening for patriotic Irishmen. Why not catch and preserve the fish at home, and get the entire benefit of the fish traffic? Will it be believed that there is probably more money value in the seas round Ireland than there is in the land itself? This is actually the case with the sea round the county of Aberdeen.[7]

A vast source of wealth lies at the very doors of the Irish people. But the harvest of an ocean teeming with life is allowed to pass into other hands. The majority of the boats which take part in the fishery at Kinsale are from the little island of Man, from Cornwall, from France, and from Scotland. The fishermen catch the fish, salt them, and carry them or send them away. While the Irish boats are diminishing in number, those of the strangers are increasing. In an East Lothian paper, published in May 1881, I find the following paragraph, under the head of Cockenzie:-.

"Departure of Boats.—In the early part of this week, a number of the boats here have left for the herring-fishery at Kinsale, in Ireland. The success attending their labours last year at that place and at Howth has induced more of them than usual to proceed thither this year."

It may not be generally known that Cockenzie is a little fishing village on the Firth of Forth, in Scotland, where the fishermen have provided themselves, at their own expense, with about fifty decked fishing-boats, each costing, with nets and gear, about 500L. With these boats they carry on their pursuits on the coast of Scotland, England, and Ireland. In 1882, they sent about thirty boats to Kinsale[8] and Howth. The profits of their fishing has been such as to enable them, with the assistance of Lord Wemyss, to build for themselves a convenient harbour at Port Seaton, without any help from the Government. They find that self-help is the best help, and that it is absurd to look to the Government and the public purse for what they can best do for themselves.

The wealth of the ocean round Ireland has long been known. As long ago as the ninth and tenth centuries, the Danes established a fishery off the western coasts, and carried on a lucrative trade with the south of Europe. In Queen Mary's reign, Philip II. of Spain paid 1000L. annually in consideration of his subjects being allowed to fish on the north-west coast of Ireland; and it appears that the money was brought into the Irish Exchequer. In 1650, Sweden was permitted, as a favour, to employ a hundred vessels in the Irish fishery; and the Dutch in the reign of Charles I. were admitted to the fisheries on the payment of 30,000L. In 1673, Sir W. Temple, in a letter to Lord Essex, says that "the fishing of Ireland might prove a mine under water as rich as any under ground."[9]

The coasts of Ireland abound in all the kinds of fish in common use—cod, ling, haddock, hake, mackerel, herring, whiting, conger, turbot, brill, bream, soles, plaice, dories, and salmon. The banks off the coast of Galway are frequented by myriads of excellent fish; yet, of the small quantity caught, the bulk is taken in the immediate neighbourhood of the shores. Galway bay is said to be the finest fishing ground in the world; but the fish cannot be expected to come on shore unsought: they must be found, followed, and netted. The fishing-boats from the west of Scotland are very successful; and they often return the fish to Ireland, cured, which had been taken out of the Irish bays. "I tested this fact in Galway," says Mr. S. C. Hall. "I had ordered fish for dinner; two salt haddocks were brought to me. On inquiry, I ascertained where they were bought, and learned from the seller that he was the agent of a Scotch firm, whose boats were at that time loading in the bay."[10] But although Scotland imports some 80,000 barrels of cured herrings annually into Ireland, that is not enough; for we find that there is a regular importation of cured herrings, cod, ling, and hake, from Newfoundland and Nova Scotia, towards the food of the Irish people.[11]

The fishing village of Claddagh, at Galway, is more decaying than ever. It seems to have suffered from a bombardment, like the rest of the town. The houses of the fishermen, when they fall in, are left in ruins. While the French, and English, and Scotch boats leave the coast laden with fish, the Claddagh men remain empty-handed. They will only fish on "lucky days," so that the Galway market is often destitute of fish, while the Claddagh people are starving. On one occasion an English company was formed for the purpose of fishing and curing fish at Galway, as is now done at Yarmouth, Grimsby, Fraserburgh, Wick, and other places. Operations were commenced, but so soon as the English fishermen put to sea in their boats, the Claddagh men fell upon them, and they were glad to escape with their lives.[12] Unfortunately, the Claddagh men have no organization, no fixed rules, no settled determination to work, unless when pressed by necessity. The appearance of the men and of their cabins show that they are greatly in want of capital; and fishing cannot be successfully performed without a sufficiency of this industrial element.

Illustrations of this neglected industry might be given to any extent. Herring fishing, cod fishing, and pilchard fishing, are alike untouched. The Irish have a strong prejudice against the pilchard; they believe it to be an unlucky fish, and that it will rot the net that takes it. The Cornishmen do not think so, for they find the pilchard fishing to be a source of great wealth. The pilchards strike upon the Irish coast first before they reach Cornwall. When Mr. Brady, Inspector of Irish Fisheries, visited St. Ives a few years ago, he saw captured, in one seine alone, nearly ten thousand pounds of this fish.

Not long since; according to a northern local paper,[13] a large fleet of vessels in full sail was seen from the west coast of Donegal, evidently making for the shore. Many surmises were made about the unusual sight. Some thought it was the Fenians, others the Home Rulers, others the Irish-American Dynamiters. Nothing of the kind! It was only a fleet of Scotch smacks, sixty-four in number, fishing for herring between Torry Island and Horn Head. The Irish might say to the Scotch fishermen, in the words of the Morayshire legend, "Rejoice, O my brethren, in the gifts of the sea, for they enrich you without making any one else the poorer!"

But while the Irish are overlooking their treasure of herring, the Scotch are carefully cultivating it. The Irish fleet of fishing-boats fell off from 27,142 in 1823 to 7181 in 1878; and in 1882 they were still further reduced to 6089.[14] Yet Ireland has a coast-line of fishing ground of nearly three thousand miles in extent.

The bights and bays on the west coast of Ireland—off Erris, Mayo, Connemara, and Donegal—swarm with fish. Near Achill Bay, 2000 mackerel were lately taken at a single haul; and Clew Bay is often alive with fish. In Scull Bay and Crookhaven, near Cape Clear, they are so plentiful that the peasants often knock them on the head with oars, but will not take the trouble to net them.

These swarms of fish might be a source of permanent wealth. A gentleman of Cork one day borrowed a common rod and line from a Cornish miner in his employment, and caught fifty-seven mackerel from the jetty in Scull Bay before breakfast. Each of these mackerel was worth twopence in Cork market, thirty miles off. Yet the people round about, many of whom were short of food, were doing nothing to catch them, but expecting Providence to supply their wants. Providence, however, always likes to be helped. Some people forget that the Giver of all good gifts requires us to seek for them by industry, prudence, and perseverance.[15]

Some cry for more loans; some cry for more harbours. It would be well to help with suitable harbours, but the system of dependence upon Government loans is pernicious. The Irish ought to feel that the very best help must come from themselves. This is the best method for teaching independence. Look at the little Isle of Man. The fishermen there never ask for loans. They look to their nets and their boats; they sail for Ireland, catch the fish, and sell them to the Irish people. With them, industry brings capital, and forms the fertile seed-ground of further increase of boats and nets. Surely what is done by the Manxmen, the Cornishmen, and the Cockenziemen, might be done by the Irishmen. The difficulty is not to be got over by lamenting about it, or by staring at it, but by grappling with it, and overcoming it. It is deeds, not words, that are wanted. Employment for the mass of the people must spring from the people themselves. Provided there is security for life and property, and an absence of intimidation, we believe that capital will become invested in the fishing industry of Ireland; and that the result will be peace, food, and prosperity.

We must remember that it is only of comparatively late years that England and Scotland have devoted so much attention to the fishery of the seas surrounding our island. In this fact there is consolation and hope for Ireland. At the beginning of the seventeenth century Sir Waiter Raleigh laid before the King his observations concerning the trade and commerce of England, in which he showed that the Dutch were almost monopolising the fishing trade, and consequently adding to their shipping, commerce, and wealth. "Surely," he says, "the stream is necessary to be turned to the good of this kingdom, to whose sea-coasts alone God has sent us these great blessings and immense riches for us to take; and that every nation should carry away out of this kingdom yearly great masses of money for fish taken in our seas, and sold again by them to us, must needs be a great dishonour to our nation, and hindrance to this realm."

The Hollanders then had about 50,000 people employed in fishing along the English coast; and their industry and enterprise gave employment to about 150,000 more, "by sea and land, to make provision, to dress and transport the fish they take, and return commodities; whereby they are enabled yearly to build 1000 ships and vessels." The prosperity of Amsterdam was then so great that it was said that Amsterdam was "founded on herring-bones." Tobias Gentleman published in 1614 his treatise on 'England's Way to win Wealth, and to employ Ships and Marines,'[16] in which he urged the English people to vie with the Dutch in fishing the seas, and thereby to give abundant employment, as well as abundant food, to the poorer people of the country.

"Look," he said, "on these fellows, that we call the plump Hollanders; behold their diligence in fishing, and our own careless negligence!" The Dutch not only fished along the coasts near Yarmouth, but their fishing vessels went north as far as the coasts of Shetland. What most roused Mr. Gentleman's indignation was, that the Dutchmen caught the fish and sold them to the Yarmouth herring-mongers "for ready gold, so that it amounteth to a great sum of money, which money doth never come again into England." "We are daily scorned," he says, "by these Hollanders, for being so negligent of our Profit, and careless of our Fishing; and they do daily flout us that be the poor Fishermen of England, to our Faces at Sea, calling to us, and saying, 'Ya English, ya sall or oud scoue dragien;' which, in English, is this, 'You English, we will make you glad to wear our old Shoes!'"

Another pamphlet, to a similar effect, 'The Royal Fishing revived,'[17] was published fifty years later, in which it was set forward that the Dutch "have not only gained to themselves almost the sole fishing in his Majesty's Seas; but principally upon this Account have very near beat us out of all our other most profitable Trades in all Parts of the World." It was even proposed to compel "all Sorts of begging Persons and all other poor People, all People condemned for less Crimes than Blood," as well as "all Persons in Prison for Debt," to take part in this fishing trade! But this was not the true way to force the traffic. The herring fishery at Yarmouth and along the coast began to make gradual progress with the growth of wealth and enterprise throughout the country; though it was not until 1787—less than a hundred years ago—that the Yarmouth men began the deep-sea herring fishery.

Before then, the fishing was all carried on along shore in little cobles, almost within sight of land. The native fishery also extended northward, along the east coast of Scotland and the Orkney and Shetland Isles, until now the herring fishery of Scotland forms one of the greatest industries in the United Kingdom, and gives employment, directly or indirectly, to close upon half a million of people, or to one-seventh of the whole population of Scotland.

Taking these facts into consideration, therefore, there is no reason to despair of seeing, before many years have elapsed, a large development of the fishing industry of Ireland. We may yet see Galway the Yarmouth, Achill the Grimsby, and Killybegs the Wick of the West. Modern society in Ireland, as everywhere else, can only be transformed through the agency of labour, industry, and commerce—inspired by the spirit of work, and maintained by the accumulations of capital. The first end of all labour is security,—security to person, possession, and property, so that all may enjoy in peace the fruits of their industry. For no liberty, no freedom, can really exist which does not include the first liberty of all—the right of public and private safety.

To show what energy and industry can do in Ireland, it is only necessary to point to Belfast, one of the most prosperous and enterprising towns in the British Islands. The land is the same, the climate is the same, and the laws are the same, as those which prevail in other parts of Ireland. Belfast is the great centre of Irish manufactures and commerce, and what she has been able to do might be done elsewhere, with the same amount of energy and enterprise. But it is not land, or climate, or altered laws that are wanted. It is men to lead and direct, and men to follow with anxious and persevering industry. It is always the Man society wants.

The influence of Belfast extends far out into the country. As you approach it from Sligo, you begin to see that you are nearing a place where industry has accumulated capital, and where it has been invested in cultivating and beautifying the land. After you pass Enniskillen, the fields become more highly cultivated. The drill-rows are more regular; the hedges are clipped; the weeds no longer hide the crops, as they sometimes do in the far west. The country is also adorned with copses, woods, and avenues. A new crop begins to appear in the fields—a crop almost peculiar to the neighbourhood of Belfast. It is a plant with a very slender erect green stem, which, when full grown, branches at the top into a loose corymb of blue flowers. This is the flax plant, the cultivation and preparation of which gives employment to a great number of persons, and is to a large extent the foundation of the prosperity of Belfast.

The first appearance of the linen industry of Ireland, as we approach Belfast from the west, is observed at Portadown. Its position on the Bann, with its water power, has enabled this town, as well as the other places on the river, to secure and maintain their due share in the linen manufacture. Factories with their long chimneys begin to appear. The fields are richly cultivated, and a general air of well-being pervades the district. Lurgan is reached, so celebrated for its diapers; and the fields there about are used as bleaching-greens. Then comes Lisburn, a populous and thriving town, the inhabitants of which are mostly engaged in their staple trade, the manufacture of damasks. This was really the first centre of the linen trade. Though Lord Strafford, during his government of Ireland, encouraged the flax industry, by sending to Holland for flax-seed, and inviting Flemish and French artisans to settle in Ireland, it was not until the Huguenots, who had been banished from France by the persecutions of Louis XIV., settled in Ireland in such large numbers, that the manufacture became firmly established. The Crommelins, the Goyers, and the Dupres, were the real founders of this great branch of industry.[18]

As the traveller approaches Belfast, groups of houses, factories, and works of various kinds, appear closer and closer; long chimneys over boilers and steam-engines, and brick buildings three or four stories high; large yards full of workmen, carts, and lorries; and at length we are landed in the midst of a large manufacturing town. As we enter the streets, everybody seems to be alive. What struck William Hutton when he first saw Birmingham, might be said of Belfast: "I was surprised at the place, but more at the people. They possessed a vivacity I had never before beheld. I had been among dreamers, but now I saw men awake. Their very step along the street showed alacrity. Every man seemed to know what he was about. The town was large, and full of inhabitants, and these inhabitants full of industry. The faces of other men seemed tinctured with an idle gloom; but here with a pleasing alertness. Their appearance was strongly marked with the modes of civil life."

Some people do not like manufacturing towns: they prefer old castles and ruins. They will find plenty of these in other parts of Ireland. But to found industries that give employment to large numbers of persons, and enable them to maintain themselves and families upon the fruits of their labour—instead of living upon poor-rates levied from the labours of others, or who are forced, by want of employment, to banish themselves from their own country, to emigrate and settle among strangers, where they know not what may become of them—is a most honourable and important source of influence, and worthy of every encouragement.

Look at the wonderfully rapid rise of Belfast, originating in the enterprise of individuals, and developed by the earnest and anxious industry of the inhabitants of Ulster!

"God save Ireland!" By all means. But Ireland cannot be saved without the help of the people who live in it. God endowed men, there as elsewhere, with reason, will, and physical power; and it is by patient industry only that they can open up a pathway to the enduring prosperity of the country. There is no Eden in nature. The earth might have continued a rude uncultivated wilderness, but for human energy, power, and industry. These enable man to subdue the wilderness, and develop the potency of labour. "Possunt quia credunt posse." They must conquer who will.

Belfast is a comparatively modern town. It has no ancient history. About the beginning of the sixteenth century it was little better than a fishing village. There was a castle, and a ford to it across the Lagan. A chapel was built at the ford, at which hurried prayers were offered up for those who were about to cross the currents of Lagan Water. In 1575, Sir Henry Sydney writes to the Lords of the Council: "I was offered skirmish by MacNeill Bryan Ertaugh at my passage over the water at Belfast, which I caused to be answered, and passed over without losse of man or horse; yet by reason of the extraordinaire Retorne our horses swamme and the Footmen in the passage waded very deep." The country round about was forest land. It was so thickly wooded that it was a common saying that one might walk to Lurgan "on the tops of the trees."

In 1612, Belfast consisted of about 120 houses, built of mud and covered with thatch. The whole value of the land on which the town is built, is said to have been worth only 5L. in fee simple.[19] "Ulster," said Sir John Davies, "is a very desert or wilderness; the inhabitants thereof having for the most part no certain habitation in any towns or villages." In 1659, Belfast contained only 600 inhabitants: Carrickfergus was more important, and had 1312 inhabitants. But about 1660, the Long Bridge over the Lagan was built, and prosperity began to dawn upon the little town. It was situated at the head of a navigable lough, and formed an outlet for the manufacturing products of the inland country. Ships of any burden, however, could not come near the town. The cargoes, down even to a recent date, had to be discharged into lighters at Garmoyle. Streams of water made their way to the Lough through the mud banks; and a rivulet ran through what is now known as the High Street.

The population gradually increased. In 1788 Belfast had 12,000 inhabitants. But it was not until after the Union with Great Britain that the town made so great a stride. At the beginning of the present century it had about 20,000 inhabitants. At every successive census, the progress made was extraordinary, until now the population of Belfast amounts to over 225,000. There is scarcely an instance of so large a rate of increase in the British Islands, save in the exceptional case of Middlesborough, which was the result of the opening out of the Stockton and Darlington Railway, and the discovery of ironstone in the hills of Cleveland in Yorkshire. Dundee and Barrow are supposed to present the next most rapid increases of population.

The increase of shipping has also been equally great. Ships from other ports frequented the Lough for purposes of trade; but in course of time the Belfast merchants supplied themselves with ships of their own. In 1791 one William Ritchie, a sturdy North Briton, brought with him from Glasgow ten men and a quantity of shipbuilding materials. He gradually increased the number of his workmen, and proceeded to build a few sloops. He reclaimed some land from the sea, and made a shipyard and graving dock on what was known as Corporation Ground. In November 1800 the new graving dock, near the bridge, was opened for the reception of vessels. It was capable of receiving three vessels of 200 tons each! In 1807 a vessel of 400 tons burthen was launched from Mr. Ritchie's shipyard, when a great crowd of people assembled to witness the launching of "so large a ship"—far more than now assemble to see a 3000-tonner of the White Star Line leave the slips and enter the water!

The shipbuilding trade has been one of the most rapidly developed, especially of late years. In 1805 the number of vessels frequenting the port was 840; whereas in 1883 the number had been increased to 7508, with about a million and a-half of tonnage; while the gross value of the exports from Belfast exceeded twenty millions sterling annually. In 1819 the first steamboat of 100 tons was used to tug the vessels up the windings of the Lough, which it did at the rate of three miles an hour, to the astonishment of everybody. Seven years later, the steamboat Rob Roy was put on between Glasgow and Belfast. But these vessels had been built in Scotland. It was not until 1826 that the first steamboat, the chieftain, was built in Belfast, by the same William Ritchie. Then, in 1838, the first iron boat was built in the Lagan foundry, by Messrs. Coates and Young, though it was but a mere cockle-shell compared with the mighty ocean steamers which are now regularly launched from Queen's Island. In the year 1883 the largest shipbuilding firm in the town launched thirteen vessels, of over 30,000 tons gross, while two other firms launched twelve ships, of about 10,000 tons gross.

I do not propose to enter into details respecting the progress of the trades of Belfast. The most important is the spinning of fine linen yarn, which is for the most part concentrated in that town, over 25,000,000 of pounds weight being exported annually. Towards the end of the seventeenth century the linen manufacture had made but little progress. In 1680 all Ireland did not export more than 6000L. worth annually. Drogheda was then of greater importance than Belfast. But with the settlement of the persecuted Hugnenots in Ulster, and especially through the energetic labours of Crommelin, Goyer, and others, the growth of flax was sedulously cultivated, and its manufacture into linen of all sorts became an important branch of Irish industry. In the course of about fifty years the exports of linen fabrics increased to the value of over 600,000L. per annum.

It was still, however, a handicraft manufacture, and done for the most part at home. Flax was spun and yarn was woven by hand. Eventually machinery was employed, and the turn-out became proportionately large and valuable. It would not be possible for hand labour to supply the amount of linen now turned out by the aid of machinery. It would require three times the entire population of Ireland to spin and weave, by the old spinning-wheel and hand-loom methods, the amount of linen cloth now annually manufactured by the operatives of Belfast alone. There are now forty large spinning-mills in Belfast and the neighbourhood, which furnish employment to a very large number of working people.[20]

In the course of my visit to Belfast, I inspected the works of the York Street flax-spinning mills, founded in 1830 by the Messrs. Mulholland, which now give employment, directly or indirectly, to many thousand persons. I visited also, with my young Italian friend, the admirable printing establishment of Marcus Ward and Co., the works of the Belfast Rope-work Company, and the shipbuilding works of Harland and Wolff. There we passed through the roar of the iron forge, the clang of the Nasmyth hammer, and the intermittent glare of the furnaces—all telling of the novel appliances of modern shipbuilding, and the power of the modern steam-engine. I prefer to give a brief account of this latter undertaking, as it exhibits one of the newest and most important industries of Belfast. It also shows, on the part of its proprietors, a brave encounter with difficulties, and sets before the friends of Ireland the truest and surest method of not only giving employment to its people, but of building up on the surest foundations the prosperity of the country.

The first occasion on which I visited Belfast—the reader will excuse the introduction of myself—was in 1840; about forty-four years ago. I went thither on the invitation of the late Wm. Sharman Crawford, Esq., M.P., the first prominent advocate of tenant-right, to attend a public meeting of the Ulster Association, and to spend a few days with him at his residence at Crawfordsburn, near Bangor. Belfast was then a town of comparatively little importance, though it had already made a fair start in commerce and industry. As our steamer approached the head of the Lough, a large number of labourers were observed—with barrows, picks, and spades—scooping out and wheeling up the slob and mud of the estuary, for the purpose of forming what is now known as Queen's Island, on the eastern side of the river Lagan. The work was conducted by William Dargan, the famous Irish contractor; and its object was to make a straight artificial outlet—the Victoria Channel—by means of which vessels drawing twenty-three feet of water might reach the port of Belfast. Before then, the course of the Lagan was tortuous and difficult of navigation; but by the straight cut, which was completed in 1846, and afterwards extended further seawards, ships of large burden were enabled to reach the quays, which extend for about a mile below Queen's Bridge, on both sides of the river.

It was a saying of honest William Dargan, that "when a thing is put anyway right at all, it takes a vast deal of mismanagement to make it go wrong." He had another curious saying about "the calf eating the cow's belly," which, he said, was not right, "at all, at all." Belfast illustrated his proverbial remarks. That the cutting of the Victoria Channel was doing the "right thing" for Belfast, was clear, from the constantly increasing traffic of the port. In course of time, several extensive docks and tidal basins were added; while provision was made, in laying out the reclaimed land at the entrance of the estuary, for their future extension and enlargement. The town of Belfast was by these means gradually placed in immediate connection by sea with the principal western ports of England and Scotland,—steamships of large burden now leaving it daily for Liverpool, Glasgow, Fleetwood, Barrow, and Ardrossan. The ships entering the port of Belfast in 1883 were 7508, of 1,526,535 tonnage; they had been more than doubled in fifteen years. The town has risen from nothing, to exhibit a Customs revenue, in 1883, of 608,781L., infinitely greater than that of Leith, the port of Edinburgh, or of Hull, the chief port of Yorkshire. The population has also largely increased. When I visited Belfast in 1840, the town contained 75,000 inhabitants. They are now over 225,006, or more than trebled,—Belfast being the tenth town, in point of population, in the United Kingdom.

The spirit and enterprise of the people are illustrated by the variety of their occupations. They do not confine themselves to one branch of business; but their energies overflow into nearly every department of industry. Their linen manufacture is of world-wide fame; but much less known are their more recent enterprises. The production of aerated waters, for instance, is something extraordinary. In 1882 the manufacturers shipped off 53,163 packages, and 24,263 cwts. of aerated waters to England, Scotland, Australia, New Zealand, and other countries. While Ireland produces no wrought iron, though it contains plenty of iron-stone,—and Belfast has to import all the iron which it consumes,—yet one engineering firm alone, that of Combe, Barbour, and Combe, employs 1500 highly-paid mechanics, and ships off its iron machinery to all parts of the world. The printing establishment of Marcus Ward and Co. employs over 1000 highly skilled and ingenious persons, and extends the influence of learning and literature into all civilised countries. We might add the various manufactures of roofing felt (of which there are five), of ropes, of stoves, of stable fittings, of nails, of starch, of machinery; all of which have earned a world-wide reputation.

We prefer, however, to give an account of the last new industry of Belfast—that of shipping and shipbuilding. Although, as we have said, Belfast imports from Scotland and England all its iron and all its coal,[21] it nevertheless, by the skill and strength of its men, sends out some of the finest and largest steamships which navigate the Atlantic and Pacific. It all comes from the power of individuality, and furnishes a splendid example for Dublin, Cork, Waterford, and Limerick, each of which is provided by nature with magnificent harbours, with fewer of those difficulties of access which Belfast has triumphed over; and each of which might be the centre of some great industrial enterprise, provided only there were patriotic men willing to embark their capital, perfect protection for the property invested, and men willing to work rather than to strike.

It was not until the year 1853 that the Queen's Island—raked out of the mud of the slob-land—was first used for shipbuilding purposes. Robert Hickson and Co. then commenced operations by laying down the Mary Stenhouse, a wooden sailing-ship of 1289 tons register; and the vessel was launched in the following year.

The operations of the firm were continued until the year 1859, when the shipbuilding establishments on Queen's Island were acquired by Mr. E. J. Harland (afterwards Harland and Wolff), since which time the development of this great branch of industry in Belfast has been rapid and complete.

From the history of this firm, it will be found that energy is the most profitable of all merchandise; and that the fruit of active work is the sweetest of all fruits. Harland and Wolff are the true Watt and Boulton of Belfast. At the beginning of their great enterprise, their works occupied about four acres of land; they now occupy over thirty-six acres. The firm has imported not less than two hundred thousand tons of iron; which have been converted by skill and labour into 168 ships of 253,000 total tonnage. These ships, if laid close together, would measure nearly eight miles in length.

The advantage to the wage-earning class can only be shortly stated. Not less than 34 per cent. is paid in labour on the cost of the ships turned out. The number of persons employed in the works is 3920; and the weekly wages paid to them is 4000L., or over 200,000L. annually. Since the commencement of the undertaking, about two millions sterling have been paid in wages.

All this goes towards the support of the various industries of the place. That the working classes of Belfast are thrifty and frugal may be inferred from the fact that at the end of 1882 they held deposits in the Savings Bank to the amount of 230,289L., besides 158,064L. in the Post Office Savings Banks.[22] Nearly all the better class working people of the town live in separate dwellings, either rented or their own property. There are ten Building Societies in Belfast, in which industrious people may store their earnings, and in course of time either buy or build their own houses.

The example of energetic, active men always spreads. Belfast contains two other shipbuilding yards, both the outcome of Harland and Wolff's enterprise; those of Messrs. Macilwaine and Lewis, employing about four hundred men, and of Messrs. Workman and Clarke, employing about a thousand. The heads of both these firms were trained in the parent shipbuilding works of Belfast. There is do feeling of rivalry between the firms, but all work together for the good of the town.

In Plutarch's Lives, we are told that Themistocles said on one occasion, "'Tis true that I have never learned how to tune a harp, or play upon a lute, but I know how to raise a small and inconsiderable city to glory and greatness." So might it be said of Harland and Wolff. They have given Belfast not only a potency for good, but a world-wide reputation. Their energies overflow. Mr. Harland is the active and ever-prudent Chairman of the most important of the local boards, the Harbour Trust of Belfast, and exerts himself to promote the extension of the harbour facilities of the port as if the benefits were to be exclusively his own; while Mr. Wolff is the Chairman of one of the latest born industries of the place, the Belfast Rope-work Company, which already gives employment to over 600 persons.

This last-mentioned industry is only about six years old. The works occupy over seven acres of ground, more than six acres of which are under roofing. Although the whole of the raw material is imported from abroad from Russia, the Philippine Islands, New Zealand, and Central America—it is exported again in a manufactured state to all parts of the world.

Such is the contagion of example, and such the ever-branching industries with which men of enterprise and industry can enrich and bless their country. The following brief memoir of the career of Mr. Harland has been furnished at my solicitation; and I think that it will be found full of interest as well as instruction.

Footnotes for Chapter X.

[1] Report in the Cork Examiner, 5th July, 1883.

[2] In 1883, as compared with 1882, there was a decrease of 58,022 acres in the land devoted to the growth of wheat; there was a total decrease of 114,871 acres in the land under tillage.—Agricultural Statistics, Ireland, 1883. Parliamentary Return, c. 3768.

[3] Statistical Abstract for the United Kingdom, 1883.

[4] The particulars are these: deposits in Irish Post Office Savings Banks, 31st December, 1882, 1,925,440; to the credit of depositors and Government stock, 125,000L.; together, 2,050,440L.

The increase of deposits over those made in the preceding year, were: in Dublin, 31,321L.; in Antrim, 23,328L.; in Tyrone, 21,315L.; in Cork, 17,034L.; and in Down, 10,382L.

[5] The only thriving manufacture now in Dublin is that of intoxicating drinks—beer, porter, stout, and whisky. Brewing and distilling do not require skilled labour, so that strikes do not affect them.

[6] Times, 11th June, 1883.

[7] The valuation of the county of Aberdeen (exclusive of the city) was recently 866,816L., whereas the value of the herrings (748,726 barrels) caught round the coast (at 25s. the barrel) was 935,907L., thereby exceeding the estimated annual rental of the county by 69,091L. The Scotch fishermen catch over a million barrels of herrings annually, representing a value of about a million and a-half sterling.

[8] A recent number of Land and Water supplies the following information as to the fishing at Kinsale:—"The takes of fish have been so enormous and unprecedented that buyers can scarcely be found, even when, as now, mackerel are selling at one shilling per six score. Piles of magnificent fish lie rotting in the sun. The sides of Kinsale Harbour

are strewn with them, and frequently, when they have become a little 'touched,' whole boat-loads are thrown overboard into the water. This great waste is to be attributed to scarcity of hands to salt the fish and want of packing-boxes. Some of the boats are said to have made as much as 500L. this season. The local fishing company are making active preparations for the approaching herring fishery, and it is anticipated that Kinsale may become one of the centres of this description of fishing."

[9] Statistical Journal for March 1848. Paper by Richard Valpy on "The Resources of the Irish Sea Fisheries," pp. 55-72.

[10] HALL, Retrospect of a Long Life, ii. 324.

[11] The Commissioners of Irish Fisheries, in one of their reports, observe:—"Notwithstanding the diminished population, the fish captured round the coast is so inadequate to the wants of the population that fully 150,000L. worth of ling, cod, and herring are annually imported from Norway, Newfoundland, and Scotland, the vessels bearing these cargoes, as they approach the shores of Ireland, frequently sailing through large shoals of fish of the same description as they are freighted with!"

[12] The following examination of Mr. J. Ennis, chairman of the Midland and Great Western Railway, took place before the "Royal Commission on Railways," as long ago as the year 1846:—

Chairman—"Is the fish traffic of any importance to your railway?"

Mr. Ennis—"of course it is, and we give it all the facilities that we can.... But the Galway fisheries, where one would expect to find plenty of fish, are totally neglected."

Sir Rowland Hill—"What is the reason of that?"

Mr. Ennis—"I will endeavour to explain. I had occasion a few nights ago to speak to a gentleman in the House of Commons with regard to an application to the Fishery Board for 2000L. to restore the pier at Buffin, in Clew Bay, and I said, 'Will you join me in the application? I am told it is a place that swarms with fish, and if we had a pier there the fishermen will have some security, and they will go out.' The only answer I received was, 'They will not go out; they pay no attention whatever to the fisheries; they allow the fish to come and go without making any effort to catch them....'"

Mr. Ayrton—"Do you think that if English fishermen went to the west coast of Ireland they would be able to get on in harmony with the native fishermen?"

Mr. Ennis—"We know the fact to be, that some years ago, a company was established for the purpose of trawling in Galway Bay, and what was the consequence? The Irish fishermen, who inhabit a region in the neighbourhood of Galway, called Claddagh, turned out against them, and would not allow them to trawl, and the Englishmen very properly went away with their lives."

Sir Rowland Hill—"Then they will neither fish themselves nor allow any one else to fish!"

Mr. Ennis—"It seems to be so."—Minutes of Evidence, 175-6.

[13] The Derry Journal.

[14] Report of Inspectors of Irish Fisheries for 1882.

[15] The Report of the Inspectors of Irish Fisheries on the Sea and Inland Fisheries of Ireland for 1882, gives a large amount of information as to the fish which swarm round the Irish coast. Mr. Brady reports on the abundance of herring and other fish all round the coast. Shoals of herrings "remained off nearly the entire coast of Ireland from August till December." "Large shoals of pilchards" were observed on the south and south-west coasts. Off Dingle, it is remarked, "the supply of all kinds of fish is practically inexhaustible."

"Immense shoals of herrings off Liscannor and Loop Head;" "the mackerel is always on this coast, and can be captured at any time of the year, weather permitting." At Belmullet, "the shoals of fish off the coast, particularly herring and mackerel, are sometimes enormous." The fishermen, though poor, are all very orderly and well conducted. They only want energy and industry.

[16] The Harleian Miscellany, iii. 378-91.

[17] The Harleian Miscellany, iii. 392.

[18] See The Huguenots in England and Ireland. A Board of Traders, for the encouragement and promotion of the hemp and flax manufacture in Ireland, was appointed by an Act of Parliament at the beginning of last century (6th October, 1711), and the year after the appointment of the Board the following notice was placed on the records of the institution:—"Louis Crommelin and the Huguenot colony have been greatly instrumental in improving and propagating the flaxen manufacture in the north of this Kingdom, and the perfection to which the same is brought in that part of the country has been greatly owing to the skill and industry of the said Crommelin." In a history of the linen trade, published at Belfast, it is said that "the dignity which that enterprising man imparted to labour, and the halo which his example cast around physical exertion, had the best effect in raising the tone of popular feeling, as well among the patricians as among the peasants of the north of Ireland. This love of industry did much to break down the national prejudice in favour of idleness, and cast doubts on the social orthodoxy of the idea then so popular with the squirearchy, that those alone who were able to live without employment had any rightful claim to the distinctive title of gentleman.... A patrician by birth and a merchant by profession, Crommelin proved, by his own life, his example, and his enterprise, that an energetic manufacturer may, at the same time, take a high place in the conventional world."

[19] Benn's History of Belfast, p. 78.

[20] From the Irish Manufacturers' Almanack for 1883 I learn that nearly one-third of the spindles used in Europe in the linen trade, and more than one-fourth of the power-looms, belong to Ireland, that "the Irish linen and associated trades at present give employment to 176,303 persons; and it is estimated that the capital sunk in spinning and weaving factories, and the business incidental thereto, is about 100,000,000L., and of that sum 37,000,000L. is credited to Belfast alone."

[21] The importation of coal in 1883 amounted to over 700,000 tons.

[22] We are indebted to the obliging kindness of the Right Hon. Mr. Fawcett, Postmaster-General for this return. The total number of depositors in the Post Office Savings banks in the Parliamentary borough of Belfast is 10,827 and the amount of their deposits, including the interest standing to their credit, on the 31st December, 1882, was 158,064L. 0s. 1d.

An important item in the savings of Belfast, not included in the above returns, consists in the amounts of deposits made with the various Limited Companies, as well as with the thriving Building Societies in the town and neighbourhood.

CHAPTER XI.

SHIPBUILDING IN BELFAST—ITS ORIGIN AND PROGRESS.

BY SIR E. J. HARLAND, ENGINEER AND SHIPBUILDER.

"The useful arts are but reproductions or new combinations by the art of man, of the same natural benefactors. He no longer waits for favouring gales, but by means of steam he realises the fable of AEolus's bag, and carries the two-and-thirty winds in the boiler of his boat."—Emerson.

"The most exquisite and the most expensive machinery is brought into play where operations on the most common materials are to be performed, because these are executed on the widest scale. This is the meaning of the vast and astonishing prevalence of machine work in this country: that the machine, with its million fingers, works for millions of purchasers, while in remote countries, where magnificence and savagery stand side by side, tens of thousands work for one. There Art labours for the rich alone; here she works for the poor no less. There the multitude produce only to give splendour and grace to the despot or the warrior, whose slaves they are, and whom they enrich; here the man who is powerful in the weapons of peace, capital, and machinery, uses them to give comfort and enjoyment to the public, whose servant he is, and thus becomes rich while he enriches others with his goods."—William Whewell, D.D.

I was born at Scarborough in May, 1831, the sixth of a family of eight. My father was a native of Rosedale, half-way between Whitby and Pickering: his nurse was the sister of Captain Scoresby, celebrated as an Arctic explorer. Arrived at manhood, he studied medicine, graduated at Edinburgh, and practised in Scarborough until nearly his death in 1866. He was thrice Mayor and a Justice of the Peace for the borough. Dr. Harland was a man of much force of character, and displayed great originality in the treatment of disease. Besides exercising skill in his profession, he had a great love for mechanical pursuits. He spent his leisure time in inventions of many sorts; and, in conjunction with the late Sir George Cayley of Brompton, he kept an excellent mechanic constantly at work.

In 1827 he invented and patented a steam-carriage for running on common roads. Before the adoption of railways, the old stage coaches were found slow and insufficient for the traffic. A working model of the steam-coach was perfected, embracing a multitubular boiler for quickly raising high-pressure steam, with a revolving surface condenser for reducing the steam to water again, by means of its exposure to the cold draught of the atmosphere through the interstices of extremely thin laminations of copper plates. The entire machinery, placed under the bottom of the carriage, was borne on springs; the whole being of an elegant form. This model steam-carriage ascended with perfect ease the steepest roads. Its success was so complete that Dr. Harland designed a full-sized carriage; but the demands upon his professional skill were so great that he was prevented going further than constructing the pair of engines, the wheels, and a part of the boiler,—all of which remnants I still preserve, as valuable links in the progress of steam locomotion.

Other branches of practical science—such as electricity, magnetism, and chemical cultivation of the soil—received a share of his attention. He predicted that three or four powerful electric lamps would yet light a whole city. He was also convinced of the feasibility of an electric cable to New York, and calculated the probable cost. As an example to the neighbourhood, he successfully cultivated a tract of moorland, and overcame difficulties which before then were thought insurmountable.

When passing through Newcastle, while still a young man, on one of his journeys to the University at Edinburgh, and being desirous of witnessing the operations in a coal-mine, a friend recommended him to visit Killingworth pit, where he would find one George Stephenson, a most intelligent workman, in charge. My father was introduced to Mr. Stephenson accordingly; and after rambling over the underground workings, and observing the pumping and winding engines in full operation, a friendship was made, which afterwards proved of the greatest service to myself, by facilitating my being placed as a pupil at the great engineering works of Messrs. Robert Stephenson and Co., at Newcastle.

My mother was the daughter of Gawan Pierson, a landed proprietor of Goathland, near Rosedale. She, too, was surprisingly mechanical in her tastes; and assisted my father in preparing many of his plans, besides attaining considerable proficiency in drawing, painting, and modelling in wax. Toys in those days were poor, as well as very expensive to purchase. But the nursery soon became a little workshop under her directions; and the boys were usually engaged, one in making a cart, another in carving out a horse, and a third in cutting out a boat; while the girls were making harness, or sewing sails, or cutting out and making perfect dresses for their dolls—whose houses were completely furnished with everything, from the kitchen to the attic, all made at home.

It was in a house of such industry and mechanism that I was brought up. As a youth, I was slow at my lessons; preferring to watch and assist workmen when I had an opportunity of doing so, even with the certainty of having a thrashing from the schoolmaster for my neglect. Thus I got to know every workshop and every workman in the town. At any rate I picked up a smattering of a variety of trades, which afterwards proved of the greatest use to me. The chief of these was wooden shipbuilding, a branch of industry then extensively carried on by Messrs. William and Robert Tindall, the former of whom resided in London; he was one of the half-dozen great shipbuilders and owners who founded "Lloyd's." Splendid East Indiamen, of some 1000 tons burden, were then built at Scarborough; and scarcely a timber was moulded, a plank bent, a spar lined off, or launching ship-ways laid, without my being present to witness them. And thus, in course of time, I was able to make for myself the neatest and fastest of model yachts.

At that time, I attended the Grammar School. Of the rudiments taught, I was fondest of drawing, geometry, and Euclid. Indeed, I went twice through the first two books of the latter before I was twelve years old. At this age I was sent to the Edinburgh Academy, my eldest brother William being then a medical student at the University. I remained at Edinburgh two years. My early progress in mathematics would have been lost in the classical training which was then insisted upon at the academy, but for my brother who was not only a good mathematician but an excellent mechanic. He took care to carry on my instruction in that branch of knowledge, as well as to teach me to make models of machines and buildings, in which he was himself proficient. I remember, in one of my journeys to Edinburgh, by coach from Darlington, that a gentleman expressed his wonder what a screw propeller could be like; for the screw, as a method of propulsion, was then being introduced. I pointed out to him the patent tail of a windmill by the roadside, and said, "It is just like that!"

In 1844 my mother died; and shortly after, my brother having become M.D., and obtained a prize gold medal, we returned to Scarborough. It was intended that he should assist my father; but he preferred going abroad for a few years. I may mention further, with relation to him, that after many years of scientific research and professional practice, he died at Hong Kong in 1858, when a public monument was erected to his memory, in what is known as the "Happy Valley."

I remained for a short time under the tuition of my old master. But as the time was rapidly approaching when I too must determine what I was "to be" in life. I had no hesitation in deciding to be an engineer, though my father wished me to be a barrister. But I kept constant to my resolution; and eventually he succeeded, through his early acquaintance with George Stephenson, in gaining for me an entrance to the engineering works of Robert Stephenson and

Co., at Newcastle-upon-Tyne. I started there as a pupil on my fifteenth birthday, for an apprenticeship of five years. I was to spend the first four years in the various workshops, and the last year in the drawing-office.

I was now in my element. The working hours, it is true, were very long,—being from six in the morning until 8.15 at night; excepting on Saturday, when we knocked off at four. However, all this gave me so much the more experience; and, taking advantage of it, I found that, when I had reached the age of eighteen, I was intrusted with the full charge of erecting one side of a locomotive. I had to accomplish the same amount of work as my mate on the other side, one Murray Playfair, a powerful, hard-working Scotchman. My strength and endurance were sometimes taxed to the utmost, and required the intervals of my labour to be spent in merely eating and sleeping.

I afterwards went through the machine-shops. I was fortunate enough to get charge of the best screw-cutting and brass-turning lathe in the shop; the former occupant, Jack Singleton, having just been promoted to a foreman's berth at the Messrs. Armstrong's factory. He afterwards became superintendent of all the hydraulic machinery of the Mersey Dock Trust at Liverpool. After my four years had been completed, I went into the drawing-office, to which I had looked forward with pleasure; and, having before practised lineal as well as free-hand drawing, I soon succeeded in getting good and difficult designs to work out, and eventually finished drawings of the engines. Indeed, on visiting the works many years after, one of these drawings was shown to me as a "specimen;" the person exhibiting it not knowing that it was my own work.

In the course of my occasional visits to Scarborough, my attention was drawn to the imperfect design of the lifeboats of the period; the frequent shipwrecks along the coast indicating the necessity for their improvement. After considerable deliberation, I matured a plan for a metal lifeboat, of a cylindrico-conical or chrysalis form, to be propelled by a screw at each end, turned by sixteen men inside, seated on water-ballast tanks; sufficient room being left at the ends inside for the accommodation of ten or twelve shipwrecked persons; while a mate near the bow, and the captain near the stern in charge of the rudder, were stationed in recesses in the deck about three feet deep. The whole apparatus was almost cylindrical, and watertight, save in the self-acting ventilators, which could only give access to the smallest portion of water. I considered that, if the lifeboat fully manned were launched into the roughest seas, or off the deck of a vessel, it would, even if turned on its back, immediately right itself, without any of the crew being disturbed from their positions, to which they were to have been strapped.

It happened that at this time (the summer of 1850) his Grace the late Duke of Northumberland, who had always taken a deep interest in the Lifeboat Institution, offered a prize of one hundred guineas for the best model and design of such a craft; so I determined to complete my plans and make a working model of my lifeboat. I came to the conclusion that the cylindrico-conical form, with the frames to be carried completely round and forming beams as well, and the two screws, one at each end, worked off the same power, by which one or other of them would always be immersed, were worth registering in the Patent Office. I therefore entered a caveat there; and continued working at my model in the evenings. I first made a wooden block model, on the scale of an inch to the foot. I had some difficulty in procuring sheets of copper thin enough, so that the model should draw only the correct amount of water; but at last I succeeded, through finding the man at Newcastle who had supplied my father with copper plates for his early road locomotive.

The model was only 32 inches in length, and 8 inches in beam; and in order to fix all the internal fittings, of tanks, seats, crank handles, and pulleys, I had first to fit the shell plating, and then, by finally securing one strake of plates on, and then another, after all inside was complete, I at last finished for good the last outside plate. In executing the job, my early experience of all sorts of handiwork came serviceably to my aid. After many a whole night's work—for the evenings alone were not sufficient for the purpose—I at length completed my model; and triumphantly and confidently took it to sea in an open boat; and then cast it into the waves. The model either rode over them or passed through them; if it was sometimes rolled over, it righted itself at once, and resumed its proper attitude in the waters. After a considerable trial I found scarcely a trace of water inside. Such as had got there was merely through the joints in the sliding hatches; though the ventilators were free to work during the experiments.

I completed the prescribed drawings and specifications, and sent them, together with the model, to Somerset House. Some 280 schemes of lifeboats were submitted for competition; but mine was not successful. I suspect that the extreme novelty of the arrangement deterred the adjudicators from awarding in its favour. Indeed, the scheme was so unprecedented, and so entirely out of the ordinary course of things, that there was no special mention made of it in the report afterwards published, and even the description there given was incorrect. The prize was awarded to Mr. James Beeching, of Great Yarmouth, whose plans were afterwards generally adopted by the Lifeboat Society. I have preserved my model just as it was; and some of its features have since been introduced with advantage into shipbuilding.[1]

The firm of Robert Stephenson and Co. having contracted to build for the Government three large iron caissons for the Keyham Docks, and as these were very similar in construction to that of an ordinary iron ship, draughtsmen conversant with that class of work were specially engaged to superintend it. The manager, knowing my fondness for ships, placed me as his assistant at this new work. After I had mastered it, I endeavoured to introduce improvements, having observed certain defects in laying down the lines—I mean by the use of graduated curves cut out of thin wood. In lieu of this method, I contrived thin tapered laths of lancewood, and weights of a particular form, with steel claws and knife edges attached, so as to hold the lath tightly down to the paper, yet capable of being readily adjusted, so as to produce any form of curve, along which the pen could freely and continuously travel. This method proved very efficient, and it has since come into general use.

The Messrs. Stephenson were then also making marine engines, as well as large condensing pumping engines, and a large tubular bridge to be erected over the river Don. The splendid high-level bridge over the Tyne, of which Robert Stephenson was the engineer, was also in course of construction. With the opportunity of seeing these great works in progress, and of visiting, during my holidays and long evenings, most of the manufactories and mines in the neighbourhood of Newcastle, I could not fail to pick up considerable knowledge, and an acquaintance with a vast variety of trades. There were about thirty other pupils in the works at the same time with myself; some were there either through favour or idle fancy; but comparatively few gave their full attention to the work, and I have since heard nothing of them. Indeed, unless a young fellow takes a real interest in his work, and has a genuine love for it, the greatest advantages will prove of no avail whatever.

It was a good plan adopted at the works, to require the pupils to keep the same hours as the rest of the men, and, though they paid a premium on entering, to give them the same rate of wages as the rest of the lads. Mr. William Hutchinson, a contemporary of George Stephenson, was the managing partner. He was a person of great experience, and had the most thorough knowledge of men and materials, knowing well how to handle both to the best advantage.

His son-in-law, Mr. William Weallans, was the head draughtsman, and very proficient, not only in quickness but in accuracy and finish. I found it of great advantage to have the benefit of the example and the training of these very clever men.

My five years apprenticeship was completed in May 1851, on my twentieth birthday. Having had but very little "black time," as it was called, beyond the half-yearly holiday for visiting my friends, and having only "slept in" twice during the five years, I was at once entered on the books as a journeyman, on the "big" wage of twenty shillings a week. Orders were, however, at that time very difficult to be had.

Railway trucks, and even navvies' barrows, were contracted for in order to keep the men employed. It was better not to discharge them, and to find something for them to do. At the same time it was not very encouraging for me, under such circumstances, to remain with the firm. I therefore soon arranged to leave; and first of all I went to see London. It was the Great Exhibition year of 1851. I need scarcely say what a rich feast I found there, and how thoroughly I enjoyed it all. I spent about two months in inspecting the works of art and mechanics in the Exhibition, to my own great advantage. I then returned home; and, after remaining in Scarborough for a short time, I proceeded to Glasgow with a letter of introduction to Messrs. J. and G. Thomson, marine engine builders, who started me on the same wages which I had received at Stephenson's, namely twenty shillings a week.

I found the banks of the Clyde splendid ground for gaining further mechanical knowledge. There were the ship and engine works on both sides of the river, down to Govan; and below there, at Renfrew, Dumbarton, Port Glasgow, and Greenock—no end of magnificent yards—so that I had plenty of occupation for my leisure time on Saturday afternoons. The works of Messrs. Robert Napier and Sons were then at the top of the tree. The largest Cunard steamers were built and engined there. Tod and Macgregor were the foremost in screw steamships—those for the Peninsular and Oriental Company being splendid models of symmetry and works of art. Some of the fine wooden paddle-steamers built in Bristol for the Royal Mail Company were sent round to the Clyde for their machinery. I contrived to board all these ships from time to time, so as to become well acquainted with their respective merits and peculiarities.

As an illustration of how contrivances, excellent in principle, but defective in construction, may be discarded, but again taken up under more favourable circumstances, I may mention that I saw a Hall's patent surface-condensor thrown to one side from one of these steamers, the principal difficulty being in keeping it tight. And yet, in the course of a very few years, by the simplest possible contrivance—inserting an indiarubber ring round each end of the tube (Spencer's patent)—surface condensation in marine engines came into vogue; and there is probably no ocean-going steamer afloat without it, furnished with every variety of suitable packings.

After some time, the Messrs. Thomson determined to build their own vessels, and an experienced naval draughtsman was engaged, to whom I was "told off" whenever he needed assistance. In the course of time, more and more of the ship work came in my way. Indeed, I seemed to obtain the preference. Fortunately for us both, my superior obtained an appointment of a similar kind on the Tyne, at superior pay, and I was promoted to his place. The Thomsons had now a very fine shipbuilding-yard, in full working order, with several large steamers on the stocks. I was placed in the drawing-office as head draughtsman. At the same time I had no rise of wages; but still went on enjoying my twenty shillings a week. I was, however, gaining information and experience, and knew that better pay would follow in due course of time. And without solicitation I was eventually offered an engagement for a term of years, at an increased and increasing salary, with three months' notice on either side.

I had only enjoyed the advance for a short time, when Mr. Thomas Toward, a shipbuilder on the Tyne, being in want of a manager, made application to the Messrs. Stephenson for such a person. They mentioned my name, and Mr. Toward came over to the Clyde to see me. The result was, that I became engaged, and it was arranged that I should enter on my enlarged duties on the Tyne in the autumn of 1853. It was with no small reluctance that I left the Messrs. Thomson. They were first-class practical men, and had throughout shown me every kindness and consideration. But a managership was not to be had every day; and being the next step to the position of a master, I could not neglect the opportunity for advancement which now offered itself.

Before leaving Glasgow, however, I found that it would be necessary to have a new angle and plate furnace provided for the works on the Tyne. Now, the best man in Glasgow for building these important requisites for shipbuilding work was scarcely ever sober; but by watching and coaxing him, and by a liberal supply of Glenlivat afterwards, I contrived to lay down on paper, from his directions, what he considered to be the best class of furnace; and by the aid of this I was afterwards enabled to construct what proved to be the best furnace on the Tyne.

To return to my education in shipbuilding. My early efforts in ship-draughting at Stephensons' were further developed and matured at Thomsons' on the Clyde. Models and drawings were more carefully worked out on the 1/4-in. scale than heretofore. The stern frames were laid off and put up at once correctly, which before had been first shaped by full-sized wooden moulds. I also contrived a mode of quickly and correctly laying off the frame-lines on a model, by laying it on a plane surface, and then, with a rectangular block traversing it—a pencil in a suitable holder being readily applied over the curved surface. This method is now in general use.

Even at that time, competition as regards speed in the Clyde steamers was very keen. Foremost among the competitors was the late Mr. David Hutchinson, who, though delighted with the Mountaineer, built by the Thomsons in 1853, did not hesitate to have her lengthened forward to make her sharper, so as to secure her ascendency in speed during the ensuing season. The results were satisfactory; and his steamers grew and grew, until they developed into the celebrated Iona and Cambria, which were in later years built for him by the same firm. I may mention that the Cunard screw steamer Jura was the last heavy job with which I was connected while at Thomsons'.

I then proceeded to the Tyne, to superintend the building of ships and marine boilers. The shipbuilding yard was at St. Peter's, about two and a-half miles below Newcastle. I found the work, as practised there, rough and ready; but by steady attention to all the details, and by careful inspection when passing the "piece-work" (a practice much in vogue there, but which I discouraged), I contrived to raise the standard of excellence, without a corresponding increase of price. My object was to raise the quality of the work turned out; and, as we had orders from the Russian Government, from China, and the Continent, as well as from shipowners at home, I observed that quality was a very important element in all commercial success. My master, Mr. Thomas Toward, was in declining health; and, being desirous of spending his winters abroad, I was consequently left in full charge of the works. But as there did not appear to be a satisfactory prospect, under the circumstances, for any material development of the business, a trifling circumstance arose, which again changed the course of my career.

An advertisement appeared in the papers for a manager to conduct a shipbuilding yard in Belfast. I made inquiries as to the situation, and eventually applied for it. I was appointed, and entered upon my duties there at Christmas, 1854. The yard was a much larger one than that on the Tyne, and was capable of great expansion. It was situated on what was then well known as the Queen's Island; but now, like the Isle of Dogs, it has been attached by reclamation. The yard, about four acres in extent, was held by lease from the Belfast Harbour Commissioners. It was well placed, alongside a fine patent slip, with clear frontage, allowing of the largest ships being freely launched. Indeed, the first ship built there, the Mary Stenhouse, had only just been completed and launched by Messrs. Robert Hickson and Co., then the proprietors of the undertaking. They were also the owners of the Eliza Street Iron Works, Belfast, which were started to work up old iron materials. But as the works were found to be unremunerative, they were shortly afterwards closed.

On my entering the shipbuilding yard I found that the firm had an order for two large sailing ships. One of these was partly in frame; and I at once tackled with it and the men. Mr. Hickson, the acting partner, not being practically acquainted with the business, the whole proceeding connected with the building of the ships devolved upon me. I had been engaged to supersede a manager summarily dismissed. Although he had not given satisfaction to his employers, he was a great favourite with the men. Accordingly, my appearance as manager in his stead was not very agreeable to the employed. On inquiry I found that the rate of wages paid was above the usual value, whilst the quantity as well as quality of the work done were below the standard. I proceeded to rectify these defects, by paying the ordinary rate of wages, and then by raising the quality of the work done. I was met by the usual method—a strike. The men turned out. They were abetted by the former manager; and the leading hands hung about the town unemployed, in the hope of my throwing up the post in disgust.

But, nothing daunted, I went repeatedly over to the Clyde for the purpose of enlisting fresh hands. When I brought them over, however, in batches, there was the greatest difficulty in inducing them to work. They were intimidated, or enticed, or feasted, and sent home again. The late manager had also taken a yard on the other side of the river, and actually commenced to build a ship, employing some of his old comrades; but beyond laying the keel, little more was ever done. A few months after my arrival, my firm had to arrange with its creditors, whilst I, pending the settlement, had myself to guarantee the wages to a few of the leading hands, whom I had only just succeeded in gathering together. In this dilemma, an old friend, a foreman on the Clyde, came over to Belfast to see me. After hearing my story, and considering the difficulties I had to encounter, he advised me at once to "throw up the job!" My reply was, that "having mounted a restive horse, I would ride him into the stable."

Notwithstanding the advice of my friend, I held on. The comparatively few men in the works, as well as those out, no doubt observed my determination. The obstacles were no doubt great; the financial difficulties were extreme; and yet there was a prospect of profit from the work in hand, provided only the men could be induced to settle steadily down to their ordinary employment. I gradually gathered together a number of steady workmen, and appointed suitable foremen. I obtained a considerable accession of strength from Newcastle. On the death of Mr. Toward, his head foreman, Mr. William Hanston, with a number of the leading hands, joined me. From that time forward the works went on apace; and we finished the ships in hand to the perfect satisfaction of the owners.

Orders were obtained for several large sailing ships as well as screw vessels. We lifted and repaired wrecked ships, to the material advantage of Mr. Hickson, then the sole representative of the firm. After three years thus engaged, I resolved to start somewhere as a shipbuilder on my own account. I made inquiries at Garston, Birkenhead, and other places. When Mr. Hickson heard of my intentions, he said he had no wish to carry on the concern after I left, and made a satisfactory proposal for the sale to me of his holding of the Queen's Island Yard. So I agreed to the proposed arrangement. The transfer and the purchase were soon completed, through the kind assistance of my old and esteemed friend Mr. G. G. Schwabe, of Liverpool; whose nephew, Mr. G. W. Wolff, had been with me for a few months as my private assistant.

It was necessary, however, before commencing for myself, that I should assist Mr. Hickson in finishing off the remaining vessels in hand, as well as to look out for orders on my own account. Fortunately, I had not long to wait; for it had so happened that my introduction to the Messrs. Thomson of Glasgow had been made through the instrumentality of my good friend Mr. Schwabe, who induced Mr. James Bibby (of J. Bibby, Sons & Co., Liverpool) to furnish me with the necessary letter. While in Glasgow, I had endeavoured to assist the Messrs. Bibby in the purchase of a steamer; so I was now intrusted by them with the building of three screw steamers the Venetian, Sicilian, and Syrian, each 270 feet long, by 34 feet beam, and 22 feet 9 inches hold; and contracted with Macnab and Co., Greenock, to supply the requisite steam-engines.

This was considered a large order in those days. It required many additions to the machinery, plant, and tools of the yard. I invited Mr. Wolff, then away in the Mediterranean as engineer of a steamer, to return and take charge of the drawing office. Mr. Wolff had served his apprenticeship with Messrs. Joseph Whitworth and Co., of Manchester, and was a most able man, thoroughly competent for the work. Everything went on prosperously; and, in the midst of all my engagements, I found time to woo and win the hand of Miss Rosa Wann, of Vermont, Belfast, to whom I was married on the 26th of January, 1860, and by her great energy, soundness of judgment, and cleverness in organization, I was soon relieved from all sources of care and anxiety, excepting those connected with business.

The steamers were completed in the course of the following year, doubtless to the satisfaction of the owners, for their delivery was immediately followed by an order for two larger vessels. As I required frequently to go from home, and as the works must be carefully attended to during my absence, on the 1st of January, 1862, I took Mr. Wolff in as a partner; and the firm has since continued under the name of Harland and Wolff. I may here add that I have throughout received the most able advice and assistance from my excellent friend and partner, and that we have together been enabled to found an entirely new branch of industry in Belfast.

It is necessary for me here to refer back a little to a screw steamer which was built on the Clyde for Bibby and Co. by Mr. John Read, and engined by J. and G. Thomson while I was with them. That steamer was called the Tiber. She was looked upon as of an extreme length, being 235 feet, in proportion to her beam, which was 29 feet. Serious misgivings were thrown out as to whether she would ever stand a heavy sea. Vessels of such proportions were thought to be crank, and even dangerous. Nevertheless, she seemed to my mind a great success. From that time, I began to think and work out the advantages and disadvantages of such a vessel, from an owner's as well as from a builder's point of view. The result was greatly in favour of the owner, though entailing difficulties in construction as regards the builder. These difficulties, however. I thought might easily be overcome.

In the first steamers ordered of me by the Messrs. Bibby, I thought it more prudent to simply build to the dimensions furnished, although they were even longer than usual. But, prior to the precise dimensions being fixed for the second order, I with confidence proposed my theory of the greater carrying power and accommodation, both for cargo and passengers, that would be gained by constructing the new vessels of increased length, without any increase of beam. I conceived that they would show improved qualities in a sea-way, and that, notwithstanding the increased accommodation, the same speed with the same power would be obtained, by only a slight increase in the first cost. The result was, that I was allowed to settle the dimensions; and the following were then decided on: Length, 310 feet; beam, 34 feet; depth of hold, 24 feet 9 inches; all of which were fully compensated for by making the upper deck entirely of iron. In this way, the hull of the ship was converted into a box girder of immensely increased strength, and was, I believe, the first ocean steamer ever so constructed. The rig too was unique. The four masts were made in one continuous length, with fore-and-aft sails, but no yards,—thereby reducing the number of hands necessary to work them. And the steam winches were so arranged as to be serviceable for all the heavy hauls, as well as for the rapid handling of the cargo.

In the introduction of so many novelties, I was well supported by Mr. F. Leyland, the junior partner of Messrs. Bibby's firm, and by the intelligent and practical experience of Captain Birch, the overlooker, and Captain George Wakeham, the Commodore of the company. Unsuccessful attempts had been made many years before to condense the steam from the engines by passing it into variously formed chambers, tubes, &c., to be there condensed by surfaces kept cold by the circulation of sea-water round them, so as to preserve the pure water and return it to the boilers free of salt. In this way, "salting up" was avoided, and a considerable saving of fuel and expenses in repairs was effected.

Mr. Spencer had patented an improvement on Hall's method of surface condensation, by introducing indiarubber rings at each end of the tubes. This had been tried as an experiment on shore, and we advised that it should be adopted in one of Messrs. Bibby's smallest steamers, the Frankfort. The results were found perfectly satisfactory. Some 20 per cent. of fuel was saved; and, after the patent right had been bought, the method was adopted in all the vessels of the company.

When these new ships were first seen at Liverpool, the "old salts" held up their hands. They were too long! they were too sharp! they would break their backs! They might, indeed, get out of the Mersey, but they would never get back! The ships, however, sailed; and they made rapid and prosperous voyages to and from the Mediterranean. They fulfilled all the promises which had been made. They proved the advantages of our new build of ships; and the owners were perfectly satisfied with their superior strength, speed, and accommodation. The Bibbys were wise men in their day and generation. They did not stop, but went on ordering more ships. After the Grecian and the Italian had made two or three voyages to Alexandria, they sent us an order for three more vessels. By our advice, they were made twenty feet longer than the previous ones, though of no greater beam; in other respects, they were almost identical. This was too much for "Jack."

"What!" he exclaimed, "more Bibby's coffins?" Yes, more and more; and in the course of time, most shipowners followed our example.

To a young firm, a repetition of orders like these was a great advantage,—not only because of the novel design of the ships, but also because of their constructive details. We did our best to fit up the Egyptian, Dalmatian, and Arabian, as first-rate vessels. Those engaged in the Mediterranean trade finding them to be serious rivals, partly because of the great cargos which they carried, but principally from the regularity with which they made their voyages with such surprisingly small consumption of coal. They were not, however, what "Jack" had been accustomed to consider "dry ships." The ship built Dutchman fashion, with her bluff ends, is the driest of all ships, but the least steady, because she rises to every sea. But the new ships, because of their length and sharpness, precluded this; for, though they rose sufficiently to an approaching wave for all purposes of safety, they often went through the crest of it, and, though shipping a little water, it was not only easier for the vessel, but the shortest road.

Nature seems to have furnished us with the finest design for a vessel in the form of the fish: it presents such fine lines—is so clean, so true, and so rapid in its movements. The ship, however, must float; and to hit upon the happy medium of velocity and stability seems to me the art and mystery of shipbuilding. In order to give large carrying capacity, we gave flatness of bottom and squareness of bilge. This became known in Liverpool as the "Belfast bottom;" and it has been generally adopted. This form not only serves to give stability, but also increases the carrying power without lessening the speed.

While Sailor Jack and our many commercial rivals stood aghast and wondered, our friends gave us yet another order for a still longer ship, with still the same beam and power. The vessel was named the Persian; she was 360 feet long, 34 feet beam, 24 feet 9 inches hold. More cargo was thus carried, at higher speed. It was only a further development of the fish form of structure. Venice was an important port to call at. The channel was difficult to navigate, and the Venetian class (270 feet long) was supposed to be the extreme length that could be handled here. But what with the straight stem,—by cutting the forefoot away, and by the introduction of powerful steering-gear, worked amidships,—the captain was able to navigate the Persian, 90 feet longer than the Venetian, with much less anxiety and inconvenience.

Until the building of the Persian, we had taken great pride in the modelling and finish of the old style of cutwater and figurehead, with bowsprit and jib-boom; but in urging the advantages of greater length of hull, we were met by the fact of its being simply impossible in certain docks to swing vessels of any greater length than those already constructed. Not to be beaten, we proposed to do away with all these overhanging encumbrances, and to adopt a perpendicular stem. In this way the hull might be made so much longer; and this was, I believe, the first occasion of its being adopted in this country in the case of an ocean steamer; though the once celebrated Collins Line of paddle steamers had, I believe, such stems. The iron decks, iron bulwarks, and iron rails, were all found very serviceable in our later vessels, there being no leaking, no caulking of deck-planks or waterways, nor any consequent damaging of cargo.

Having found it impossible to combine satisfactorily wood with iron, each being so differently affected by temperature and moisture, I secured some of these novelties of construction in a patent, by which filling in the spaces between frames, &c., with Portland cement, instead of chocks of wood, and covering the iron plates with cement and tiles, came into practice, and this has since come into very general use.

The Tiber, already referred to, was 235 feet in length when first constructed by Read, of Glasgow, and was then thought too long; but she was now placed in our hands to be lengthened 39 feet, as well as to have an iron deck added, both of which greatly improved her. We also lengthened the Messrs. Bibby's Calpe—also built by Messrs. Thomson while I was there—by no less than 93 feet. The advantage of lengthening ships, retaining the same beam and power, having become generally recognised, we were in trusted by the Cunard Company to lengthen the Hecla, Olympus, Atlas, and Marathon, each by 63 feet. The Royal Consort P.S., which had been lengthened first at Liverpool, was again lengthened by us at Belfast.

The success of all this heavy work, executed for successful owners, put a sort of backbone into the Belfast shipbuilding yard. While other concerns were slack, we were either lengthening or building steamers as well as sailing-ships for firms in Liverpool, London, and Belfast. Many acres of ground were added to the works. The Harbour Commissioners had now made a fine new graving-dock, and connected the Queen's Island with the mainland. The yard, thus improved and extended, was surveyed by the Admiralty, and placed on the first-class list. We afterwards built for the Government the gun vessels Lynx and Algerine, as well as the store and torpedo ship Hecla, of 3360 tons.

The Suez Canal being now open, our friends the Messrs. Bibby gave us an order for three steamers of very large tonnage, capable of being adapted for trade with the antipodes if necessary. In these new vessels there was no retrograde step as regards length, for they were 390 feet keel by 37 feet beam, square-rigged on three of the masts, with the yards for the first time fitted on travellers, as to enable them to be readily sent down; thus forming a unique combination of big fore-and-aft sails, with handy square sails. These ships were named the Istrian, Iberian, and Illyrian, and in 1868 they went to sea; soon after to be followed by three more ships—the Bavarian, Bohemian, and Bulgarian—in most respects the same, though ten feet longer, with the same beam. They were first placed in the Mediterranean trade, but were afterwards transferred to the Liverpool and Boston trade, for cattle and emigrants. These, with three smaller steamers for the Spanish cattle trade, and two larger steamers for other trades, made together twenty steam-vessels constructed for the Messrs. John Bibby, Sons, & Co.; and it was a matter of congratulation that, after a great deal of heavy and constant work, not one of them had exhibited the slightest indication of weakness,—all continuing in first-rate working order.

The speedy and economic working of the Belfast steamers, compared with those of the ordinary type, having now become well known, a scheme was set on foot in 1869 for employing similar vessels, though of larger size, for passenger and goods accommodation between England and America. Mr. T. H. Ismay, of Liverpool, the spirited shipowner, then formed, in conjunction with the late Mr. G. H. Fletcher, the Oceanic Steam Navigation Company, Limited; and we were commissioned by them to build six large Transatlantic steamers, capable of carrying a heavy cargo of goods, as well as a full complement of cabin and steerage passengers, between Liverpool and New York, at a speed equal, if not superior, to that of the Cunard and Inman lines. The vessels were to be longer than any we had yet constructed, being 420 feet keel and 41 feet beam, with 32 feet hold.

This was a great opportunity, and we eagerly embraced it. The works were now up to the mark in point of extent and appliances. The men in our employment were mostly of our own training: the foremen had been promoted from the ranks; the manager, Mr. W. H. Wilson, and the head draughtsman, Mr. W. J. Pirrie (since become partners), having, as pupils, worked up through all the departments, and ultimately won their honourable and responsible positions by dint of merit only—by character, perseverance, and ability. We were therefore in a position to take up an important contract of this kind, and to work it out with heart and soul.

As everything in the way of saving of fuel was of first-rate importance, we devoted ourselves to that branch of economic working. It was necessary that buoyancy or space should be left for cargo, at the same time that increased speed should be secured, with as little consumption of coal as possible. The Messrs. Elder and Co., of Glasgow, had made great strides in this direction with the paddle steam-engines which they had constructed for the Pacific Company on the compound principle. They had also introduced them on some of their screw steamers, with more or less success. Others were trying the same principle in various forms, by the use of high-pressure cylinders, and so on; the form of the boilers being varied according to circumstances, for the proper economy of fuel. The first thing absolutely wanted was, perfectly reliable information as to the actual state of the compound engine and boiler up to the date of our inquiry. To ascertain the facts by experience, we dispatched Mr. Alexander Wilson, younger brother of the manager who had been formerly a pupil of Messrs. Macnab and Co., of Greenock, and was thoroughly able for the work—to make a number of voyages in steam vessels fitted with the best examples of compound engines.

The result of this careful inquiry was the design of the machinery and boilers of the Oceanic and five sister-ships. They were constructed on the vertical overhead "tandem" type, with five-feet stroke (at that time thought excessive), oval single-ended transverse boilers, with a working pressure of sixty pounds. We contracted with Messrs. Maudslay, Sons, and Field, of London, for three of these sets, and with Messrs. George Forrester and Co., of Liverpool, for the other three; and as we found we could build the six vessels in the same time as the machinery was being constructed; and, as all this machinery had to be conveyed to Belfast to be there fitted on board, whilst the vessels were being otherwise finished, we built a little screw-steamer, the Camel, of extra strength, with very big hatchways, to receive these large masses of iron; and this, in course of time, was found to work with great advantage; until eventually we constructed our own machinery.

We were most fortunate in the type of engine we had fixed upon, for it proved both economical and serviceable in all ways; and, with but slight modifications, we repeated it in the many subsequent vessels which we built for the White Star Company. Another feature of novelty in these vessels consisted in placing the first-class accommodation amidships, with the third-class aft and forward. In all previous ocean steamers, the cabin passengers had been berthed near the stern, where the heaving motion of the vessel was far greater than in the centre, and where that most disagreeable vibration inseparable from proximity to the propeller was ever present. The unappetising smells from the galley were also avoided. And last, but not least, a commodious smoking-saloon was fitted up amidships, contrasting most favourably with the scanty accommodation provided in other vessels. The saloon, too, presented the novelty of extending the full width of the vessel, and was lighted from each side. Electric bells were for the first time fitted on board ship. The saloon and entire range of cabins were lighted by gas, made on board, though this has since given place to the incandescent electric light. A fine promenade deck was provided over the saloon, which was accessible from below in all weathers by the grand staircase.

These, and other arrangements, greatly promoted the comfort and convenience of the cabin passengers; while those in the steerage found great improvements in convenience, sanitation, and accommodation. "Jack" had his forecastle well ventilated and lighted, and a turtle-back over his head when on deck, with winches to haul for him, and a steam-engine to work the wheel; while the engineers and firemen berthed as near their work as possible, never needing to wet a jacket or miss a meal. In short, for the first time perhaps, ocean-voyaging, even in the North Atlantic, was made not only less tedious and dreadful to all, but was rendered enjoyable and even delightful to many. Before the Oceanic, the pioneer of the new line, was even launched, rival companies had already consigned her to the deepest place in the ocean. Her first appearance in Liverpool was therefore regarded with much interest. Mr. Ismay, during the construction of the vessel, took every pains to suggest improvements and arrangements with a view to the comfort and convenience of the travelling public. He accompanied the vessel on her first voyage to New York in March, 1871, under command of Captain, now Sir Digby Murray, Brt. Although severe weather was experienced, the ship made a splendid voyage, with a heavy cargo of goods and passengers. The Oceanic thus started the Transatlantic traffic of the Company, with the house-flag of the White Star proudly flying on the main.

It may be mentioned that the speed of the Oceanic was at least a knot faster per hour than had been heretofore accomplished across the Atlantic. The motion of the vessel was easy, without any indication of weakness or straining, even in the heaviest weather. The only inducement to slow was when going head to it (which often meant head through it), to avoid the inconvenience of shipping a heavy body of "green sea" on deck forward. A turtle-back was therefore provided to throw it off, which proved so satisfactory, as it had done on the Holyhead and Kingstown boats, that all the subsequent vessels were similarly constructed. Thus, then, as with the machinery, so was the hull of the Oceanic, a type of the succeeding vessels, which after intervals of a few months took up their stations on the Transatlantic line.

Having often observed, when at sea in heavy weather, how the pitching of the vessel caused the weights on the safety-valves to act irregularly, thus letting puffs of steam escape at every heave, and as high pressure steam was too valuable a commodity to be so wasted, we determined to try direct-acting spiral springs, similar to those used in locomotives, in connection with the compound engine. But as no such experiment was possible in any vessels requiring the Board of Trade certificate, the alternative of using the Camel as an experimental vessel was adopted. The spiral springs were accordingly fitted upon the boiler of that vessel, and with such a satisfactory result that the Board of Trade allowed the use of the same contrivance on all the boilers of the Oceanic and every subsequent steamer, and the contrivance has now come into general use.

It would be too tedious to mention in detail the other ships built for the White Star line. The Adriatic and Celtic were made 17 feet 6 inches longer than the Oceanic, and a little sharper, being 437 feet 6 inches keel, 41 feet beam, and 32 feet hold. The success of the Company had been so great under the able management of Ismay, Imrie and Co., and they had secured so large a share of the passengers and cargo, as well as of the mails passing between Liverpool and New York, that it was found necessary to build two still larger and faster vessels—the Britannic and Germanic: these were 455 feet in length; 45 feet in beam; and of 5000 indicated horse-power. The Britannic was in the first instance constructed with the propeller fitted to work below the line of keel when in deep water, by which means the "racing" of the engines was avoided. When approaching shallow water, the propeller was raised by steam-power to the ordinary position without any necessity for stopping the engines during the operation. Although there was an increase of speed by this means through the uniform revolutions of the machinery in the heaviest sea, yet there was an objectionable amount of vibration at certain parts of the vessel, so that we found it necessary to return to the ordinary fixed propeller, working in the line of direction of the vessel. Comfort at sea is of even more importance than speed; and although we had succeeded in four small steamers working on the new principle, it was found better to continue in the larger ships to resort to the established modes of propulsion. It may happen that at some future period the new method may yet be adopted with complete success.

Meanwhile competition went on with other companies. Monopoly cannot exist between England and America. Our plans were followed; and sharper boats and heavier power became the rule of the day. But increase of horse-power of engines means increase of heating surface and largely increased boilers, when we reach the vanishing point of profit, after which there is nothing left but speed and expense. It may be possible to fill a ship with boilers, and to save a few hours in the passage from Liverpool to New York by a tremendous expenditure of coal; but whether that will answer the purpose of any body of shareholders must be left for the future to determine.

"Brute force" may be still further employed. It is quite possible that recent "large strides" towards a more speedy transit across the Atlantic may have been made "in the dark."

The last ships we have constructed for Ismay, Imrie and Co. have been of comparatively moderate dimensions and power—the Arabic and Coptic, 430 feet long; and the Ionic and Boric,

440 feet long, all of 2700 indicated horse-power. These are large cargo steamers, with a moderate amount of saloon accommodation, and a large space for emigrants. Some of these are now engaged in crossing the Pacific, whilst others are engaged in the line from London to New Zealand; the latter being specially fitted up for carrying frozen meat.

To return to the operations of the Belfast shipbuilding yard. A serious accident occurred in the autumn of 1867 to the mail paddle-steamer the Wolf, belonging to the Messrs. Burns, of Glasgow. When passing out of the Lough, about eight miles from Belfast, she was run into by another steamer. She was cut down and sank, and there she lay in about seven fathoms of water; the top of her funnel and masts being only visible at low tide. She was in a dangerous position for all vessels navigating the entrance to the port, and it was necessary that she should be removed, either by dynamite, gunpowder, or some other process. Divers were sent down to examine the ship, and the injury done to her being found to be slight, the owners conferred with us as to the possibility of lifting her and bringing her into port. Though such a process had never before been accomplished, yet knowing her structure well, and finding that we might rely upon smooth water for about a week or two in summer, we determined to do what we could to lift the sunken vessel to the surface.

We calculated the probable weight of the vessel, and had a number of air-tanks expressly built for her floatation. These were secured to the ship with chains and hooks, the latter being inserted through the side lights in her sheer strake. Early in the following summer everything was ready. The air-tanks were prepared and rafted together. Powerful screws were attached to each chain, with hand-pumps for emptying the tanks, together with a steam tender fitted with cooking appliances, berths and stores, for all hands engaged in the enterprise. We succeeded in attaching the hooks and chains by means of divers; the chains being ready coiled on deck. But the weather, which before seemed to be settled, now gave way. No sooner had we got the pair of big tanks secured to the after body, than a fierce north-north-easterly gale set in, and we had to run for it, leaving the tanks partly filled, in order to lessen the strain on everything.

When the gale had settled, we returned again, and found that no harm had been done. The remainder of the hooks were properly attached to the rest of the tanks, the chains were screwed tightly up, and the tanks were pumped clear. Then the tide rose; and before high water we had the great satisfaction of getting the body of the vessel under weigh, and towing her about a cable's length from her old bed. At each tide's work she was lifted higher and higher, and towed into shallower water towards Belfast; until at length we had her, after eight days, safely in the harbour, ready to enter the graving dock,—not more ready, however, than we all were for our beds, for we had neither undressed nor shaved during that anxious time. Indeed, our friends scarcely recognised us on our return home.

The result of the enterprise was this. The clean cut made into the bow of the ship by the collision was soon repaired. The crop of oysters with which she was incrusted gave place to the scraper and the paintbrush. The Wolf came out of the dock to the satisfaction both of the owners and underwriters; and she was soon "ready for the road," nothing the worse for her ten months' immersion.[2]

Meanwhile the building of new iron ships went on in the Queen's Island. We were employed by another Liverpool Company—the British Shipowners' Company, Limited—to supply some large steamers. The British Empire, of 3361 gross tonnage, was the same class of vessel as those of the White Star line, but fuller, being intended for cargo. Though originally intended for the Eastern trade, this vessel was eventually placed on the Liverpool and Philadelphia line; and her working proved so satisfactory that five more vessels were ordered like her, which were chartered to the American Company.

The Liverpool agents, Messrs. Richardson, Spence, and Co., having purchased the Cunard steamer Russia, sent her over to us to be lengthened 70 feet, and entirely refitted—another proof of the rapid change which owners of merchant ships now found it necessary to adopt in view of the requirements of modern traffic.

Another Liverpool firm, the Messrs. T. and J. Brocklebank, of world-wide repute for their fine East Indiamen, having given up building for themselves at their yard at Whitehaven, commissioned us to build for them the Alexandria, and Baroda, which were shortly followed by the Candahar and Tenasserim. And continuing to have a faith in the future of big iron sailing ships, they further employed us to build for them two of yet greater tonnage, the Belfast and the Majestic.

Indeed, there is a future for sailing ships, notwithstanding the recent development of steam power. Sailing ships can still hold their own, especially in the transport of heavy merchandise for great distances. They can be built more cheaply than steamers; they can be worked more economically, because they require no expenditure on coal, nor on wages of engineers; besides, the space occupied in steamers by machinery is entirely occupied by merchandise, all of which pays its quota of freight. Another thing may be mentioned: the telegraph enables the fact of the sailing of a vessel, with its cargo on board, to be communicated from Calcutta or San Francisco to Liverpool, and from that moment the cargo becomes as marketable as if it were on the spot. There are cases, indeed, where the freight by sailing ship is even greater than by steamer, as the charge for warehousing at home is saved, and in the meantime the cargo while at sea is negotiable.

We have accordingly, during the last few years, built some of the largest iron and steel sailing ships that have ever gone to sea. The aim has been to give them great carrying capacity and fair speed, with economy of working; and the use of steel, both in the hull and the rigging, facilitates the attainment of these objects. In 1882 and 1883, we built and launched four of these steel and iron sailing ships—the Waiter H. Wilson, the W. J. Pirrie, the Fingal, and the Lord Wolseley—each of nearly 3000 tons register, with four masts,—the owners being Mr. Lawther, of Belfast; Mr. Martin, of Dublin; and the Irish Shipowners Company.

Besides these and other sailing ships, we have built for Messrs. Ismay, Imrie and Co. the Garfield, of 2347 registered tonnage; for Messrs. Thomas Dixon and Son, the Lord Downshire (2322); and for Messrs. Bullock's Bay Line, the Bay of Panama (2365).

In 1880 we took in another piece of the land reclaimed by the Belfast Harbour Trust; and there, in close proximity to the ship-yard, we manufacture all the machinery required for the service of the steamers constructed by our firm. In this way we are able to do everything "within ourselves"; and the whole land now occupied by the works comprises about forty acres, with ten building slips suitable for the largest vessels.

It remains for me to mention a Belfast firm, which has done so much for the town. I mean the Messrs. J.P. Corry and Co., who have always been amongst our best friends. We built for them their first iron sailing vessel, the Jane Porter, in 1860, and since then they have never failed us. They successfully established their "Star" line of sailing clippers from London to Calcutta, all of which were built here. They subsequently gave us orders for yet larger vessels, in the Star of France and the Star of Italy. In all, we have built for that firm eleven of their well-known "Star" ships.

We have built five ships for the Asiatic Steam Navigation Company, Limited, each of from 1650 to 2059 tons gross; and we are now building for them two ships, each of about 3000 tons gross. In 1883 we launched thirteen iron and steel vessels, of a registered tonnage of over 30,000 tons. Out of eleven ships now building, seven are of steel.

Such is a brief and summary account of the means by which we have been enabled to establish a new branch of industry in Belfast. It has been accomplished simply by energy and hard work. We have been well-supported by the skilled labour of our artisans; we have been backed by the capital and the enterprise of England; and we believe that if all true patriots would go and do likewise, there would be nothing to fear for the prosperity and success of Ireland.

Footnotes for Chapter XI.

[1] Although Mr. Harland took no further steps with his lifeboat, the project seems well worthy of a fair trial. We had lately the pleasure of seeing the model launched and tried on the lake behind Mr. Harland's residence at Ormiston, near Belfast. The cylindrical lifeboat kept perfectly water-tight, and though thrown into the water in many different positions—sometimes tumbled in on its prow, at other times on its back (the deck being undermost), it invariably righted itself. The screws fore and aft worked well, and were capable of being turned by human labour or by steam power. Now that such large freights of passengers are carried by ocean-going ships, it would seem necessary that some such method should be adopted of preserving life at sea; for ordinary lifeboats, which are so subject to destructive damage, are often of little use in fires or shipwrecks, or other accidents on the ocean.

[2] A full account is given in the Illustrated London News of the 21st of October, 1868, with illustrations, of the raising of the Wolf; and another, more scientific, is given in the Engineer of the 16th of October, of the same year.

CHAPTER XII.

ASTRONOMERS AND STUDENTS IN HUMBLE LIFE: A NEW CHAPTER IN THE 'PURSUIT OF KNOWLEDGE UNDER DIFFICULTIES.'

"I first learnt to read when the masons were at work in your house. I approached them one day, and observed that the architect used a rule and compass, and that he made calculations. I inquired what might be the meaning and use of these things, and I was informed that there was a science called Arithmetic. I purchased a book of arithmetic, and I learned it. I was told there was another science called Geometry; I bought the necessary books, and I learned Geometry. By reading, I found there were good books in these two sciences in Latin; I bought a dictionary, and I learned Latin. I understood, also, that there were good books of the same kind in French; I bought a dictionary, and I learned French. It seems to me that one does not need to know anything more than the twenty-four letters to learn everything else that one wishes."—Edmund Stone to the Duke of Argyll. ('Pursuit of Knowledge under Difficulties.')

"The British Census proper reckons twenty-seven and a half million in the home countries. What makes this census important is the quality of the units that compose it. They are free forcible men, in a country where life is safe, and has reached the greatest value. They give the bias to the current age; and that not by chance or by mass, but by their character, and by the number of individuals among them of personal ability."—Emerson: English Traits.

From Belfast to the Highlands of Scotland is an easy route by steamers and railways. While at Birnam, near Dunkeld, I was reminded of some remarkable characters in the neighbourhood. After the publication of the 'Scotch Naturalist' and 'Robert Dick,' I received numerous letters informing me of many self-taught botanists and students of nature, quite as interesting as the subjects of my memoirs. Among others, there was John Duncan, the botanist weaver of Aberdeen, whose interesting life has since been done justice to by Mr. Jolly; and John Sim of

Perth, first a shepherd boy, then a soldier, and towards the close of his life a poet and a botanist, whose life, I was told, was "as interesting as a romance."

There was also Alexander Croall, Custodian of the Smith Institute at Stirling, an admirable naturalist and botanist. He was originally a hard-working parish schoolmaster, near Montrose. During his holiday wanderings he collected plants for his extensive herbarium. His accomplishments having come under the notice of the late Sir William Hooker, he was selected by that gentleman to prepare sets of the Plants of Braemar for the Queen and Prince Albert, which he did to their entire satisfaction. He gave up his school-mastership for an ill-paid but more congenial occupation, that of Librarian to the Derby Museum and Herbarium. Some years ago, he was appointed to his present position of Custodian to the Smith Institute—perhaps the best provincial museum and art gallery in Scotland.

I could not, however, enter into the history of these remarkable persons; though I understand there is a probability of Mr. Croall giving his scientific recollections to the world. He has already brought out a beautiful work, in four volumes, 'British Seaweeds, Nature-printed;' and anything connected with his biography will be looked forward to with interest.

Among the other persons brought to my notice, years ago, were Astronomers in humble life. For instance, I received a letter from John Grierson, keeper of the Girdleness Lighthouse, near Aberdeen, mentioning one of these persons as "an extraordinary character." "William Ballingall," he said, "is a weaver in the town of Lower Largo, Fifeshire; and from his early days he has made astronomy the subject of passionate study. I used to spend my school vacation at Largo, and have frequently heard him expound upon his favourite subject. I believe that very high opinions have been expressed by scientific gentlemen regarding Ballingall's attainments. They were no doubt surprised that an individual with but a very limited amount of education, and whose hours of labour were from five in the morning until ten or eleven at night, should be able to acquire so much knowledge on so profound a subject. Had he possessed a fair amount of education, and an assortment of scientific instruments and books, the world would have heard more about him. Should you ever find yourself," my correspondent concludes, "in his neighbourhood, and have a few hours to spare, you would have no reason to regret the time spent in his company." I could not, however, arrange to pay the proposed visit to Largo; but I found that I could, without inconvenience, visit another astronomer in the neighbourhood of Dunkeld.

In January 1879 I received a letter from Sheriff Barclay, of Perth, to the following effect: "Knowing the deep interest you take in genius and merit in humble ranks, I beg to state to you an extraordinary case. John Robertson is a railway porter at Coupar Angus station. From early youth he has made the heavens his study. Night after night he looks above, and from his small earnings he has provided himself with a telescope which cost him about 30L. He sends notices of his observations to the scientific journals, under the modest initials of 'J.R.' He is a great favourite with the public; and it is said that he has made some observations in celestial phenomena not before noticed. It does occur to me that he should have a wider field for his favourite study. In connection with an observatory, his services would be invaluable."

Nearly five years had elapsed since the receipt of this letter, and I had done nothing to put myself in communication with the Coupar Angus astronomer. Strange to say, his existence was again recalled to my notice by Professor Grainger Stewart, of Edinburgh. He said that if I was in the neighbourhood I ought to call upon him, and that he would receive me kindly. His duty, he said, was to act as porter at the station, and to shout the name of the place as the trains passed. I wrote to John Robertson accordingly, and received a reply stating that he would be glad to see me, and inclosing a photograph, in which I recognised a good, honest, sensible face, with his person inclosed in the usual station porter's garb, "C.R. 1446."

I started from Dunkeld, and reached Coupar Angus in due time. As I approached the station, I heard the porter calling out, "Coupar Angus! change here for Blairgowrie!"[1] It was the voice of John Robertson.

I descended from the train, and addressed him at once: after the photograph there could be no mistaking him. An arrangement for a meeting was made, and he called upon me in the evening. I invited him to such hospitality as the inn afforded; but he would have nothing. "I am much obliged to you," he said; "but it always does me harm." I knew at once what the "it" meant. Then he invited me to his house in Causewayend Street. I found his cottage clean and comfortable, presided over by an evidently clever wife. He took me into his sitting-room, where I inspected his drawings of the sun-spots, made in colour on a large scale. In all his statements he was perfectly modest and unpretending. The following is his story, so far as I can recollect, in his own words:—

"Yes; I certainly take a great interest in astronomy, but I have done nothing in it worthy of notice. I am scarcely worthy to be called a day labourer in the science. I am very well known hereabouts, especially to the travelling public; but I must say that they think a great deal more of me than I deserve.

"What made me first devote my attention to the subject of astronomy? Well, if I can trace it to one thing more than another, it was to some evening lectures delivered by the late Dr. Dick, of Broughty Ferry, to the men employed at the Craigs' Bleachfield Works, near Montrose, where I then worked, about the year 1848. Dr. Dick was an excellent lecturer, and I listened to him with attention. His instructions were fully impressed upon our minds by Mr. Cooper, the teacher of the evening school, which I attended. After giving the young lads employed at the works their lessons in arithmetic, he would come out with us into the night—and it was generally late when we separated—and show us the principal constellations, and the planets above the horizon. It was a wonderful sight; yet we were told that these hundreds upon hundreds of stars, as far as the eye could see, were but a mere vestige of the creation amidst which we lived. I got to know the names of some of the constellations the Greater Bear, with 'the pointers' which pointed to the Pole Star, Orion with his belt, the Twins, the Pleiades, and other prominent objects in the heavens. It was a source of constant wonder and surprise.

"When I left the Bleachfield Works, I went to Inverury, to the North of Scotland Railway, which was then in course of formation; and for many years, being immersed in work, I thought comparatively little of astronomy. It remained, however, a pleasant memory. It was only after coming to this neighbourhood in 1854, when the railway to Blairgowrie was under construction, that I began to read up a little, during my leisure hours, on the subject of astronomy. I got married the year after, since which time I have lived in this house.

"I became a member of a reading-room club, and read all the works of Dr. Dick that the library contained: his 'Treatise on the Solar System,' his 'Practical Astronomer,' and other works. There were also some very good popular works to which I was indebted for amusement as well as instruction: Chambers's 'Information for the People,' Cassell's 'Popular Educator,' and a very interesting series of articles in the 'Leisure Hour,' by Edwin Dunkin of the Royal Observatory, Greenwich. These last papers were accompanied by maps of the chief constellations, so that I had a renewed opportunity of becoming a little better acquainted with the geography of the heavens.

"I began to have a wish for a telescope, by means of which I might be able to see a little more than with my naked eyes. But I found that I could not get anything of much use, short of 20L. I could not for a long time feel justified in spending so much money for my own personal enjoyment. My children were then young and dependent upon me. They required to attend school—for education is a thing that parents must not neglect, with a view to the future. However, about the year 1875, my attention was called to a cheap instrument advertised by Solomon—what he called his '5L. telescope.' I purchased one, and it tantalised me; for the power of the instrument was such as to teach me nothing of the surface of the planets. After using it for about two years, I sold it to a student, and then found that I had accumulated enough savings to enable me to buy my present instrument. Will you come into the next room and look at it?"

I went accordingly into the adjoining room, and looked at the new telescope. It was taken from its case, put upon its tripod, and looked in beautiful condition. It is a refractor, made by Cooke and Sons of York. The object glass is three inches; the focal length forty-three inches; and the telescope, when drawn out, with the pancratic eyepiece attached, is about four feet. It was made after Mr. Robertson's directions, and is a sort of combination of instruments.

"Even that instrument," he proceeded, "good as it is for the money, tantalises me yet. A look through a fixed equatorial, such as every large observatory is furnished with is a glorious view. I shall never forget the sight that I got when at Dunecht Observatory, to which I was invited through the kindness of Dr. Copeland, the Earl of Crawford and Balcarres' principal astronomer.

"You ask me what I have done in astronomical research? I am sorry to say I have been able to do little except to gratify my own curiosity; and even then, as I say, I have been much tantalised. I have watched the spots on the sun from day to day through obscured glasses, since the year 1878, and made many drawings of them. Mr. Rand Capron, the astronomer, of Guildown, Guildford, desired to see these drawings, and after expressing his satisfaction with them, he sent them to Mr. Christie, Astronomer Royal, Greenwich. Although photographs of the solar surface were preferred, Mr. Capron thought that my sketches might supply gaps in the partially cloudy days, as well as details which might not appear on the photographic plates. I received a very kind letter from Mr. Christie, in which he said that it would be very difficult to make the results obtained from drawings, however accurate, at all comparable with those derived from photographs; especially as regards the accurate size of the spots as compared with the diameter of the sun. And no doubt he is right.

"What, do I suppose, is the cause of these spots in the sun? Well, that is a very difficult question to answer. Changes are constantly going on at the sun's surface, or, I may rather say, in the sun's interior, and making themselves apparent at the surface. Sometimes they go on with enormous activity; at other times they are more quiet. They recur alternately in periods of seven or eight weeks, while these again are also subject to a period of about eleven years—that is, the short recurring outbursts go on for some years, when they attain a maximum, from which they go on decreasing. I may say that we are now (August 1883) at, or very near, a maximum epoch. There is no doubt that this period has an intimate connection with our auroral displays; but I don't think that the influence sun-spots have on light or heat is perceptible. Whatever influence they possess would be felt alike on the whole terrestrial globe. We have wet, dry, cold, and warm years, but they are never general. The kind of season which prevails in one country is often quite reversed in another perhaps in the adjacent one. Not so with our auroral displays. They are universal on both sides of the globe; and from pole to pole the magnetic needle trembles during their continuance. Some authorities are of opinion that these eleven-year cycles are subject to a larger cycle, but sun-spot observations have not existed long enough to determine this point. For myself, I have a great difficulty in forming an opinion. I have very little doubt that the spots are depressions on the surface of the sun. This is more apparent when the spot is on the limb. I have often seen the edge very rugged and uneven when groups of large spots were about to come round on the east side. I have communicated some of my observations to 'The Observatory,' the monthly review of astronomy, edited by Mr. Christie, now Astronomer Royal,[2] as well as to The Scotsmam, and some of our local papers.[3]

"I have also taken up the observation of variable stars in a limited portion of the heavens. That, and 'hunting for comets' is about all the real astronomical work that an amateur can do nowadays in our climate, with a three-inch telescope. I am greatly indebted to the Earl of Crawford and Balcarres, who regularly sends me circulars of all astronomical discoveries, both in this and foreign countries. I will give an instance of the usefulness of these circulars. On the morning of the 4th of October, 1880, a comet was discovered by Hartwig, of Strasburg, in the constellation of Corona. He telegraphed it to Dunecht Observatory, fifteen miles from Aberdeen. The circulars announcing the discovery were printed and despatched by post to various astronomers. My circular reached me by 7 P.M., and, the night being favourable, I directed my telescope upon the part of the heavens indicated, and found the comet almost at once—that is, within fifteen hours of the date of its discovery at Strasburg.

"In April, 1878, a large meteor was observed in broad daylight, passing from south to north, and falling it was supposed, about twenty miles south of Ballater. Mr. A. S. Herschel, Professor of Physics in the College of Science, 'Newcastle-on-Tyne, published a letter in The Scotsmam, intimating his desire to be informed of the particulars of the meteor's flight by those who had seen it. As I was one of those who had observed the splendid meteor flash northwards almost under the face of the bright sun (at 10.25 A.M.), I sent the Professor a full account of what I had seen, for which he professed his strong obligations. This led to a very pleasant correspondence with Professor Herschel. After this, I devoted considerable attention to meteors, and sent many contributions to 'The Observatory' on the subject.[4]

"You ask me what are the hours at which I make my observations? I am due at the railway station at six in the morning, and I leave at six in the evening; but I have two hours during the day for meals and rest. Sometimes I get a glance at the heavens in the winter mornings when the sky is clear, hunting for comets. My observations on the sun are usually made twice a day during my meal hours, or in the early morning or late at evening in summer, while the sun is visible. Yes, you are right; I try and make the best use of my time. It is much too short for all that I propose to do. My evenings are my own. When the heavens are clear, I watch them; when obscured, there are my books and letters.

"Dr. Alexander Brown, of Arbroath, is one of my correspondents. I have sent him my drawings of the rings of Saturn, of Jupiter's belt and satellites. Dr. Ralph Copeland, of Dunecht, is also a very good friend and adviser. Occasionally, too, I send accounts of solar disturbances, comet a within sight, eclipses, and occultations, to the Scotsman, the Dundee Evening Telegraph and Evening News, or to the Blairgowrie Advertiser. Besides, I am the local observer of meteorology, and communicate regularly with Mr. Symons. These things entirely fill up my time.

"Do I intend always to remain a railway porter? Oh, yes; I am very comfortable! The company are very kind to me, and I hope I serve them faithfully. It is true Sheriff Barclay has, without my knowledge, recommended me to several well-known astronomers as an observer. But at my time of life changes are not to be desired. I am quite satisfied to go on as I am doing. My young people are growing up, and are willing to work for themselves. But come, sir," he concluded, "come into the garden, and look at the moon through my telescope."

We went into the garden accordingly, but a cloud was over the moon, and we could not see it. At the top of the garden was the self-registering barometer, the pitcher to measure the rainfall, and the other apparatus necessary to enable the "Diagram of barometer, thermometer, rain, and wind" to be conducted, so far as Coupar Angus is concerned. This Mr. Robertson has done for four years past. As the hour was late, and as I knew that my entertainer must be up by six next morning, I took my leave.

A man's character often exhibits itself in his amusements. One must have a high respect for the character of John Robertson, who looks at the manner in which he spends his spare time. His astronomical work is altogether a labour of love. It is his hobby; and the working man may have his hobby as well as the rich. In his case he is never less idle than when idle. Some may think that he is casting his bread upon the waters, and that he may find it after many days. But it is not with this object that he carries on his leisure-hour pursuits. Some have tried—sheriff Barclay among others[5]—to obtain appointments for him in connection with astronomical observation; others to secure advancement for him in his own line. But he is a man who is satisfied with his lot—one of the rarest things on earth. Perhaps it is by looking so much up to the heavens that he has been enabled to obtain his portion of contentment.

Next morning I found him busy at the station, making arrangements for the departure of the passenger train for Perth, and evidently upon the best of terms with everybody. And here I leave John Robertson, the contented Coupar Angus astronomer.

Some years ago I received from my friend Mr. Nasmyth a letter of introduction to the late Mr. Cooke of York, while the latter was still living. I did not present it at the time; but I now proposed to visit, on my return homewards, the establishment which he had founded at York for the manufacture of telescopes and other optical instruments. Indeed, what a man may do for himself as well as for science, cannot be better illustrated than by the life of this remarkable man.

Mr. Nasmyth says that he had an account from Cooke himself of his small beginnings. He was originally a shoemaker in a small country village. Many a man has risen to distinction from a shoemaker's seat. Bulwer, in his 'What will He do with It?' has discussed the difference between shoemakers and tailors. "The one is thrown upon his own resources, the other works in the company of his fellows: the one thinks, the other communicates. Cooke was a man of natural ability, and he made the best use of his powers. Opportunity, sooner or later, comes to nearly all who work and wait, and are duly persevering. Shoemaking was not found very productive; and Cooke, being fairly educated as well as self-educated, opened a village school. He succeeded tolerably well. He taught himself geometry and mathematics, and daily application made him more perfect in his studies. In course of time an extraordinary ambition took possession of him: no less than the construction of a reflecting telescope of six inches diameter. The idea would not let him rest until he had accomplished his purpose. He cast and polished the speculum with great labour; but just as he was about to finish it, the casting broke! What was to be done? About one-fifth had broken away, but still there remained a large piece, which he proceeded to grind down to a proper diameter. His perseverance was rewarded by the possession of a 3 1/2 inch speculum, which by his rare skill he worked into a reflecting telescope of very good quality.

He was, however, so much annoyed by the treacherously brittle nature of the speculum metal that he abandoned its use, and betook himself to glass. He found that before he could make a good achromatic telescope it was necessary that he should calculate his curves from data depending upon the nature of the glass. He accordingly proceeded to study the optical laws of refraction, in which his knowledge of geometry and mathematics greatly helped him. And in course of time, by his rare and exquisite manipulative skill, he succeeded in constructing a four-inch refractor, or achromatic telescope, of admirable defining power.

The excellence of his first works became noised abroad. Astronomical observers took an interest in him; and friends began to gather round him, amongst others the late Professor Phillips and the Rev. Vernon Harcourt, Dean of York. Cooke received an order for a telescope like his own; then he received other orders. At last he gave up teaching, and took to telescope making. He advanced step by step; and like a practical, thoughtful man, he invented special tools and machinery for the purpose of grinding and polishing his glasses. He opened a shop in York, and established himself as a professed maker of telescopes. He added to this the business of a general optician, his wife attending to the sale in the shop, while he himself attended to the workshop.

Such was the excellence of his work that the demand for his telescopes largely increased. They were not only better manufactured, but greatly cheaper than those which had before been in common use. Three of the London makers had before possessed a monopoly of the business; but now the trade was thrown open by the enterprise of Cooke of York. He proceeded to erect a complete factory—the Buckingham Street works. His brother took charge of the grinding and polishing of the lenses, while his sons attended to the mechanism of the workshop; but Cooke himself was the master spirit of the whole concern. Everything that he did was good and accurate. His clocks were about the best that could be made. He carried out his clock-making business with the same zeal that he devoted to the perfection of his achromatic telescopes. His work was always first-rate. There was no scamping about it. Everything that he did was thoroughly good and honest. His 4 1/4-inch equatorials are perfect gems; and his admirable achromatics, many of them of the largest class, are known all over the world. Altogether, Thomas Cooke was a remarkable instance of the power of Self-Help.

Such was the story of his Life, as communicated by Mr. Nasmyth. I was afterwards enabled, through the kind assistance of his widow, Mrs. Cooke, whom I saw at Saltburn, in Yorkshire, to add a few particulars to his biography.

"My husband," she said, "was the son of a working shoemaker at Pocklington, in the East Riding. He was born in 1807. His father's circumstances were so straitened that he was not able to do much for him; but he sent him to the National school, where he received some education. He remained there for about two years, and then he was put to his father's trade. But he greatly disliked shoemaking, and longed to get away from it. He liked the sun, the sky, and the open air. He was eager to be a sailor, and, having heard of the voyages of Captain Cook, he wished to go to sea. He spent his spare hours in learning navigation, that he might be a good seaman. But when he was ready to set out for Hull, the entreaties and tears of his mother prevailed on him to give up the project; and then he had to consider what he should do to maintain himself at home.

"He proceeded with his self-education, and with such small aids as he could procure, he gathered together a good deal of knowledge. He thought that he might be able to teach others. Everybody liked him, for his diligence, his application, and his good sense. At the age of seventeen he was employed to teach the sons of the neighbouring farmers. He succeeded so well that in the following year he opened a village school at Beilby. He went on educating himself, and learnt a little of everything. He next removed his school to Kirpenbeck, near Stamford

Bridge; and it was there," proceeded Mrs. Cooke, "that I got to know him, for I was one of his pupils."

"He first learned mathematics by buying an old volume at a bookstall, with a spare shilling. That was before he began to teach. He also got odd sheets, and read other books about geometry and mathematics, before he could buy them; for he had very little to spare. He studied and learnt as much as he could.

He was very anxious to get an insight into knowledge. He studied optics before he had any teaching. Then he tried to turn his knowledge to account. While at Kirpenbeck he made his first object-glass out of a thick tumbler bottom. He ground the glass cleverly by hand; then he got a piece of tin and soldered it together, and mounted the object-glass in it so as to form a telescope.

"He next got a situation at the Rev. Mr. Shapkley's school in Micklegate, York, where he taught mathematics. He also taught in ladies' schools in the city, and did what he could to make a little income. Our intimacy had increased, and we had arranged to get married. He was twenty-four, and I was nineteen, when we were happily united. I was then his pupil for life.

"Professor Phillips saw his first telescope, with the object-glass made out of the thick tumbler bottom, and he was so much pleased with it that my husband made it over to him. But he also got an order for another, from Mr. Gray, solicitor, more by way of encouragement than because Mr. Gray wanted it, for he was a most kind man. The object-glass was of four-inch aperture, and when mounted the defining power was found excellent. My husband was so successful with his telescopes that he went on from smaller to greater, and at length he began to think of devoting himself to optics altogether. His knowledge of mathematics had led him on, and friends were always ready to encourage him in his pursuits.

"During this time he had continued his teaching at the school in the day-time; and he also taught on his own account the sons of gentlemen in the evening: amongst others the sons of Dr. Wake and Dr. Belcomb, both medical men. He was only making about 100L. a year, and his family was increasing. It was necessary to be very economical, and I was careful of everything. At length my uncle Milner agreed to advance about 100L. as a loan. A shop was taken in Stonegate in 1836, and provided with optical instruments. I attended to the shop, while my husband worked in the back premises. To bring in a little ready money, I also took in lodgers.

"My husband now devoted himself entirely to telescope making and optics. But he took in other work. His pumps were considered excellent; and he furnished all those used at the pump-room, Harrogate. His clocks, telescope-driving[6] and others, were of the best. He commenced turret-clock making in 1852, and made many improvements in them. We had by that time removed to Coney Street; and in 1855 the Buckingham Works were established, where a large number of first-rate workmen were employed. A place was also taken in Southampton Street, London, in 1868, for the sale of the instruments manufactured at York."

Thus far Mrs. Cooke. It may be added that Thomas Cooke revived the art of making refracting telescopes in England. Since the discovery by Dollond, in 1758, of the relation between the refractive and dispersive powers of different kinds of glass, and the invention by that distinguished optician of the achromatic telescope, the manufacture of that instrument had been confined to England, where the best flint glass was made. But through the short-sighted policy of the Government, an exorbitant duty was placed upon the manufacture of flint glass, and the English trade was almost entirely stamped out. We had accordingly to look to foreign countries for the further improvement of the achromatic telescope, which Dollond had so much advanced.

A humble mechanic of Brenetz, in the Canton of Neufchatel, Switzerland, named Guinaud, having directed his attention to the manufacture of flint glass towards the close of last century, at length succeeded, after persevering efforts, in producing masses of that substance perfectly free from stain, and therefore adapted for the construction of the object-glasses of telescopes.

Frauenhofer, the Bavarian optician, having just begun business, heard of the wonderful success of Guinaud, and induced the Swiss mechanic to leave Brenetz and enter into partnership with him at Munich in 1805.

The result was perfectly successful; and the new firm turned out some of the largest object-glasses which had until then been made. With one of these instruments, having an aperture of 9.9 inches, Struve, the Russian astronomer, made some of his greatest discoveries. Frauenhofer was succeeded by Merz and Mahler, who carried out his views, and turned out the famous refractors of Pulkowa Observatory in Russia, and of Harvard University in the United States. These last two telescopes contained object-glasses of fifteen inches aperture.

The pernicious impost upon flint glass having at length been removed by the English Government, an opportunity was afforded to our native opticians to recover the supremacy which they had so long lost. It is to Thomas Cooke, more than to any other person, that we owe the recovery of this manufacture. Mr. Lockyer, writing in 1878, says: "The two largest and most perfectly mounted refractors on the German form at present in existence are those at Gateshead and Washington, U.S. The former belongs to Mr. Newall, a gentleman who, connected with those who were among the first to recognise the genius of our great English optician, Cooke, did not hesitate to risk thousands of pounds in one great experiment, the success of which will have a most important bearing upon the astronomy of the future."[7]

The progress which Mr. Cooke made in his enterprise was slow but steady. Shortly after he began business as an optician, he became dissatisfied with the method of hand-polishing, and made arrangements to polish the object-glasses by machinery worked by steam power. By this means he secured perfect accuracy of figure. He was also able to turn out a large quantity of glasses, so as to furnish astronomers in all parts of the world with telescopes of admirable defining power, at a comparatively moderate price. In all his works he endeavoured to introduce simplicity. He left his mark on nearly every astronomical instrument. He found the equatorial comparatively clumsy; he left it nearly perfect. His beautiful "dividing machine," for marking divisions on the circles, four feet in diameter and altogether self-acting—which divides to five minutes and reads off to five seconds is not the least of his triumphs.

The following are some of his more important achromatic telescopes. In 1850, when he had been fourteen years in business, he furnished his earliest patron, Professor Phillips, with an equatorial telescope of 6 1/4 inches aperture. His second (of 6 1/8) was supplied two years later, to James Wigglesworth of Wakefield. William Gray, Solicitor, of York, one of his earliest friends, bought a 6 1/2-inch telescope in 1853. In the following year, Professor Pritchard of Oxford was supplied with a 6 1/2-inch. The other important instruments were as follows: in 1854, Dr. Fisher, Liverpool, 6 inches; in 1855, H. L. Patterson, Gateshead, 7 1/4 inches; in 1858, J. G. Barclay, Layton, Essex, 7 1/4 inches; in 1857, Isaac Fletcher, Cockermouth, 9 1/4 inches; in 1858, Sir W. Keith Murray, Ochtertyre, Crieff, 9 inches; in 1859, Captain Jacob, 9 inches; in 1860, James Nasmyth, Penshurst, 8 inches; in 1861, another telescope to J. G. Barclay, 10 inches; in 1864, the Rev. W. R. Dawes, Haddenham, Berks, 8 inches; and in 1867, Edward Crossley, Bermerside, Halifax, 9 3/8 inches.

In 1855 Mr. Cooke obtained a silver medal at the first Paris Exhibition for a six-inch equatorial telescope.[8] This was the highest prize awarded. A few years later he was invited to Osborne by the late Prince Albert, to discuss with his Royal Highness the particulars of an equatorial mounting with a clock movement, for which he subsequently received the order. On its completion he superintended the erection of the telescope, and had the honour of directing it to several of the celestial objects for the Queen and the Princess Alice, and answered their many interesting questions as to the stars and planets within sight.

Mr. Cooke was put to his mettle towards the close of his life. A contest had long prevailed among telescope makers as to who should turn out the largest refracting instrument. The two telescopes of fifteen inches aperture, prepared by Merz and Mahler, of Munich, were the largest then in existence. Their size was thought quite extraordinary. But in 1846, Mr. Alvan Clark, of Cambridgeport, Massachusetts, U.S., spent his leisure hour's in constructing small telescopes.[9] He was not an optician, nor a mathematician, but a portrait painter. He possessed, however, enough knowledge of optics and of mechanics, to enable him to make and judge a telescope. He spent some ten years in grinding lenses, and was at length enabled to produce objectives equal in quality to any ever made.

In 1853, the Rev. W. E. Dawes—one of Mr. Cooke's customers—purchased an object-glass from Mr. Clark. It was so satisfactory that he ordered several others, and finally an entire telescope. The American artist then began to be appreciated in his own country. In 1860 he received an order for a refractor of eighteen inches aperture, three inches greater than the largest which had up to that time been made. This telescope was intended for the Observatory of Mississippi; but the Civil War prevented its being removed to the South; and the telescope was sold to the Astronomical Society of Chicago and mounted in the Observatory of that city.

And now comes in the rivalry of Mr. Cooke of York, or rather of his patron, Mr. Newall of Gateshead. At the Great Exhibition of London, in 1862, two large circular blocks of glass, about two inches thick and twenty-six inches in diameter, were shown by the manufacturers, Messrs. Chance of Birmingham. These discs were found to be of perfect quality, and suitable for object-glasses of the best kind. At the close of the Exhibition, they were purchased by Mr. Newall, and transferred to the workshops of Messrs. Cooke and Sons at York. To grind and polish and mount these discs was found a work of great labour and difficulty. Mr. Lockyer says, "such an achievement marks an epoch in telescopic astronomy, and the skill of Mr. Cooke and the munificence of Mr. Newall will long be remembered."

When finished, the object-glass had an aperture of nearly twenty-five inches, and was of much greater power than the eighteen-inch Chicago instrument. The length of the tube was about thirty-two feet. The cast-iron pillar supporting the whole was nineteen feet in height from the ground, and the weight of the whole instrument was about six tons. In preparing this telescope, nearly everything, from its extraordinary size, had to be specially arranged.[10] The great anxiety involved in these arrangements, and the constant study and application told heavily upon Mr. Cooke, and though the instrument wanted only a few touches to make it complete, his health broke down, and he died on the 19th of October, 1868, at the comparatively early age of sixty-two.

Mr. Cooke's death was felt, in a measure, to be a national loss. His science and skill had restored to England the prominent position she had held in the time of Dollond; and, had he lived, even more might have been expected from him. We believe that the Gold Medal and Fellowship of the Royal Society were waiting for him; but, as one of his friends said to his widow, "neither worth nor talent avails when the great ordeal is presented to us." In a letter from Professor Pritchard, he said: "Your husband has left his mark upon his age. No optician of modern times has gained a higher reputation; and I for one do not hesitate to call his loss national; for he cannot be replaced at present by any one else in his own peculiar line. I shall carry the recollection of the affectionate esteem in which I held Thomas Cooke with me to my grave. Alas! that he should be cut off just at the moment when he was about to reap the rewards due to his unrivalled excellence. I have said that F.R.S. and medals were to be his. But he is, we fondly trust, in a better and higher state than that of earthly distinction. Best assured, your husband's name must ever be associated with the really great men of his day. Those who knew him will ever cherish his memory."

Mr. Cooke left behind him the great works which he founded in Buckingham Street, York. They still give employment to a large number of skilled and intelligent artizans. There I found many important works in progress,—the manufacture of theodolites, of prismatic compasses (for surveying), of Bolton's range finder, and of telescopes above all. In the factory yard was the commencement of the Observatory for Greenwich, to contain the late Mr. Lassell's splendid two feet Newtonian reflecting telescope, which has been presented to the nation. Mr. Cooke's spirit still haunts the works, which are carried on with the skill, the vigour, and the perseverance, transmitted by him to his sons.

While at York, I was informed by Mr. Wigglesworth, the partner of Messrs. Cooke, of an energetic young astronomer at Bainbridge, in the mountain-district of Yorkshire, who had not only been able to make a telescope of his own, but was an excellent photographer. He was not yet thirty years of age, but had encountered and conquered many difficulties. This is a sort of character which is more often to be met with in remote country places than in thickly-peopled cities. In the country a man is more of an individual; in a city he is only one of a multitude. The country boy has to rely upon himself, and has to work in comparative solitude, while the city boy is distracted by excitements. Life in the country is full of practical teachings; whereas life in the city may be degraded by frivolities and pleasures, which are too often the foes of work. Hence we have usually to go to out-of-the-way corners of the country for our hardest brain-workers. Contact with the earth is a great restorer of power; and it is to the country folks that we must ever look for the recuperative power of the nation as regards health, vigour, and manliness.

Bainbridge is a remote country village, situated among the high lands or Fells on the north-western border of Yorkshire. The mountains there send out great projecting buttresses into the dales; and the waters rush down from the hills, and form waterfalls or Forces, which Turner has done so much to illustrate. The river Bain runs into the Yore at Bainbridge, which is supposed to be the site of an old Roman station. Over the door of the Grammar School is a mermaid, said to have been found in a camp on the top of Addleborough, a remarkable limestone hill which rises to the south-east of Bainbridge. It is in this grammar-school that we find the subject of this little autobiography. He must be allowed to tell the story of his life—which he describes as 'Work: Good, Bad, and Indifferent—in his own words:

"I was born on November 20th, 1853. In my childhood I suffered from ill-health. My parents let me play about in the open air, and did not put me to school until I had turned my sixth year. One day, playing in the shoemaker's shop, William Farrel asked me if I knew my letters. I answered 'No.' He then took down a primer from a shelf, and began to teach me the alphabet, at the same time amusing me by likening the letters to familiar objects in his shop. I soon learned to read, and in about six weeks I surprised my father by reading from an easy book which the shoemaker had given me.

"My father then took me into the school, of which he was master, and my education may be said fairly to have begun. My progress, however, was very slow partly owing to ill-health, but more, I must acknowledge, to carelessness and inattention. In fact, during the first four years I was at school, I learnt very little of anything, with the exception of reciting verses, which I seemed to learn without any mental effort. My memory became very retentive. I found that by attentively reading half a page of print, or more, from any of the school-books, I could repeat the whole of it without missing a word. I can scarcely explain how I did it; but I think it was by paying strict attention to the words as words, and forming a mental picture of the paragraphs as they were grouped in the book. Certain, I am, that their sense never made much impression on me, for, when questioned by the teacher, I was always sent to the bottom of the class, though apparently I had learned my exercise to perfection.

"When I was twelve years old, I made the acquaintance of a very ingenious boy, who came to our school. Samuel Bridge was a born mechanic. Though only a year older than myself, such was his ability in the use of tools, that he could construct a model of any machine that he saw. He awakened in me a love of mechanical construction, and together we made models of colliery winding-frames, iron-rolling mills, trip-hammers, and water-wheels. Some of them were not mere toys, but constructed to scale, and were really good working models. This love of mechanical construction has never left me, and I shall always remember with affection Samuel Bridge, who first taught me to use the hammer and file. The last I heard of him was in 1875, when he passed his examination as a schoolmaster, in honours, and was at the head of his list.

"During the next two years, when between twelve and fourteen, I made comparatively slow progress at school. I remember having to write out the fourth commandment from memory. The teacher counted twenty-three mistakes in ten lines of my writing. It will be seen from this, that, as regards learning, I continued heedless and backward. About this time, my father, who was a good violinist, took me under his tuition. He made me practice on the violin about an hour and a half a day. I continued this for a long time. But the result was failure. I hated the violin, and would never play unless compelled to do so. I suppose the secret was that I had no 'ear.'

"It was different with subjects more to my mind. Looking over my father's books one day, I came upon Gregory's 'Handbook of Inorganic Chemistry,' and began reading it. I was fascinated with the book, and studied it morning, noon, and night—in fact, every time when I could snatch a few minutes. I really believe that at one time I could have repeated the whole of the book from memory. Now I found the value of arithmetic, and set to work in earnest on proportion, vulgar and decimal fractions, and, in fact, everything in school work that I could turn to account in the science of chemistry. The result of this sudden application was that I was seized with an illness. For some months I had incessant headache; my hair became dried up, then turned grey, and finally came off. Weighing myself shortly after my recovery, at the age of fifteen, I found that I just balanced fifty-six pounds. I took up mensuration, then astronomy, working at them slowly, but giving the bulk of my spare time to chemistry.

"In the year 1869, when I was sixteen years old, I came across Cuthbert Bede's book, entitled 'Photographic Pleasures.' It is an amusing book, giving an account of the rise and progress of photography, and at the same time having a good-natured laugh at it. I read the book carefully, and took up photography as an amusement, using some apparatus which belonged to my father, who had at one time dabbled in the art. I was soon able to take fair photographs. I then decided to try photography as a business. I was apprenticed to a photographer, and spent four years with him—one year at Northallerton, and three at Darlington. When my employer removed to Darlington, I joined the School of Art there.

"Having read an account of the experiments of M. E. Becquerel, a French savant, on photographing in the colours of nature, my curiosity was awakened. I carefully repeated his experiments, and convinced myself that he was correct. I continued my experiments in heliochromy for a period of about two years, during which time I made many photographs in colours, and discovered a method of developing the coloured image, which enabled me to shorten the exposure to one-fortieth of the previously-required time. During these experiments, I came upon some curious results, which, I think, might puzzle our scientific men to account for. For instance, I proved the existence of black light, or rays of such a nature as to turn the rose-coloured surface of the sensitive-plate black—that is, rays reflected from the black paint of drapery, produced black in the picture, and not the effect of darkness. I was, like Becquerel, unable to fix the coloured image without destroying the colours; though the plates would keep a long while in the dark, and could be examined in a subdued, though not in a strong light. The coloured image was faint, but the colours came out with great truth and delicacy.

"I began to attend the School of Art at Darlington on the 6th of March, 1872. I found, on attempting to draw, that I had naturally a correct eye and hand; and I made such progress, that when the students' drawings were examined, previously to sending them up to South Kensington, all my work was approved. I was then set to draw from the cast in chalk, although I had only been at the school for a month. I tried for all the four subjects at the May examination, and was fortunate enough to pass three of them, and obtained as a prize Packett's 'Sciography.' I worked hard during the next year, and sent up seventeen works; for one of these, the 'Venus de Milo,' I gained a studentship.

"I then commenced the study of human anatomy, and began water-colour painting, reading all the works upon art on which I could lay my hand. At the May examination of 1873, I completed my second-grade certificate, and at the end of the year of my studentship, I accepted the office of teacher in the School of Art. This art-training created in me a sort of disgust for photography, as I saw that the science of photography had really very little genuine art in it, and was more allied to a mechanical pursuit than to an artistic one. Now, when I look back on my past ideas, I clearly see that a great deal of this disgust was due to my ignorance and self-conceit.

"In 1874, I commenced painting in tempora, and then in oil, copying the pictures lent to the school from the South Kensington Art Library. I worked also from still life, and began sketching from nature in oil and water-colours, sometimes selling my work to help me to buy materials for art-work and scientific experiments. I was, however, able to do very little in the following year, as I was at home suffering from sciatica. For nine months I could not stand erect, but had to hobble about with a stick. This illness caused me to give up my teachership.

"Early in 1876 I returned to Darlington. I went on with my art studies and the science of chemistry; though I went no further in heliochromy. I pushed forward with anatomy. I sent about fifteen works to South Kensington, and gained as my third-grade prize in list A the 'Dictionary of Terms used in Art' by Thomas Fairholt, which I found a very useful work. Towards the end of the year, my father, whose health was declining, sent for me home to assist him in the school. I now commenced the study of Algebra and Euclid in good earnest, but found it tough work. My father, though a fair mathematician, was unable to give me any instruction; for he had been seized with paralysis, from which he never recovered. Before he died, he recommended me to try for a schoolmaster's certificate; and I promised him that I would. I obtained a situation as master of a small village school, not under Government inspection; and I studied during the year, and obtained a second class certificate at the Durham Diocesan College at Christmas, 1877. Early in the following year, the school was placed under Government inspection, and became a little more remunerative.

"I now went on with chemical analysis, making my own apparatus. Requiring an intense heat on a small scale, I invented a furnace that burnt petroleum oil. It was blown by compressed air. After many failures, I eventually succeeded in bringing it to such perfection that in 7 1/2 minutes it would bring four ounces of steel into a perfectly liquefied state. I next commenced the study of electricity and magnetism; and then acoustics, light, and heat. I constructed all my apparatus myself, and acquired the art of glass-blowing, in order to make my own chemical apparatus, and thus save expense.

"I then went on with Algebra and Euclid, and took up plane trigonometry; but I devoted most of my time to electricity and magnetism. I constructed various scientific apparatus—a syren, telephones, microphones, an Edison's megaphone, as well as an electrometer, and a machine for covering electric wire with cotton or silk. A friend having lent me a work on artificial memory, I began to study it; but the work led me into nothing but confusion, and I soon found that if I did not give it up, I should be left with no memory at all. I still went an sketching from Nature, not so much as a study, but as a means of recruiting my health, which was far from being good. At the beginning of 1881 I obtained my present situation as assistant master at the Yorebridge Grammar School, of which the Rev. W. Balderston, M.A., is principal.

"Soon after I became settled here, I spent some of my leisure time in reading Emerson's 'Optics,' a work I bought at an old bookstall. I was not very successful with it, owing to my deficient mathematical knowledge. On the May Science Examinations of 1881 taking place at Newcastle-on-Tyne, applied for permission to sit, and obtained four tickets for the following subjects:—Mathematics, Electricity and Magnetism, Acoustics, Light and Heat, and Physiography. During the preceding month I had read up the first three subjects, but, being pressed for time, I gave up the idea of taking physiography. However, on the last night of the examinations, I had some conversation with one of the students as to the subjects required for physiography. He said, 'You want a little knowledge of everything in a scientific way, and nothing much of anything.' I determined to try, for 'nothing much of anything' suited me exactly. I rose early next morning, and as soon as the shops were open I went and bought a book on the subject, 'Outlines of Physiography,' by W. Lawson, F.R.G.S. I read it all day, and at night sat for the examination. The results of my examinations were, failure in mathematics, but second class advanced grade certificates in all the others. I do not attach any credit to passing in physiography, but merely relate the circumstance as curiously showing what can be done by a good 'cram.'

"The failure in mathematics caused me to take the subject 'by the horns,' to see what I could do with it. I began by going over quadratic equations, and I gradually solved the whole of those given in Todhunter's larger 'Algebra.' Then I re-read the progressions, permutations, combinations; the binomial theorem, with indices and surds; the logarithmic theorem and series, converging and diverging. I got Todhunter's larger 'Plane Trigonometry,' and read it, with the theorems contained in it; then his 'Spherical Trigonometry;' his 'Analytical Geometry, of Two Dimensions,' and 'Conics.' I next obtained De Morgan's 'Differential and Integral Calculus,' then Woolhouse's, and lastly, Todhunter's. I found this department of mathematics difficult and perplexing to the last degree; but I mastered it sufficiently to turn it to some account. This last mathematical course represents eighteen months of hard work, and I often sat up the whole night through. One result of the application was a permanent injury to my sight.

"Wanting some object on which to apply my newly-acquired mathematical knowledge, I determined to construct an astronomical telescope. I got Airy's 'Geometrical Optics,' and read it through. Then I searched through all my English Mechanic (a scientific paper that I take), and prepared for my work by reading all the literature on the subject that I could obtain. I bought two discs of glass, of 6 1/2 inches diameter, and began to grind them to a spherical curve 12 feet radius. I got them hollowed out, but failed in fining them through lack of skill. This occurred six times in succession; but at the seventh time the polish came up beautifully, with scarcely a scratch upon the surface. Stopping my work one night, and it being starlight, I thought I would try the mirror on a star. I had a wooden frame ready for the purpose, which the carpenter had made for me. Judge of my surprise and delight when I found that the star disc enlarged nearly in the same manner from each side of the focal point, thus making it extremely probable that I had accidentally hit on a near approach to the parabola in the curve of my mirror. And such proved to be the case. I have the mirror still, and its performance is very good indeed.

"I went no further with this mirror, for fear or spoiling it. It is very slightly grey in the centre, but not sufficiently so as to materially injure its performance. I mounted it in a wooden tube, placed it on a wooden stand, and used it for a time thus mounted; but getting disgusted with the tremor and inconvenience I had to put up with, I resolved to construct for it an iron equatorial stand. I made my patterns, got them cast, turned and fitted them myself, grinding all the working parts together with emery and oil, and fitted a tangent-screw motion to drive the instrument in right ascension. Now I found the instrument a pleasure to use; and I determined to add to it divided circles, and to accurately adjust it to the meridian. I made my circles of well-seasoned mahogany, with slips of paper on their edges, dividing them with my drawing instruments, and varnishing them to keep out the wet. I shall never forget that sunny afternoon upon which I computed the hour-angle for Jupiter, and set the instrument so that by calculation Jupiter should pass through the field of the instrument at 1h. 25m. 15s. With my watch in my hand, and my eye to the eye-piece, I waited for the orb. When his glorious face appeared, almost in a direct line for the centre of the field, I could not contain my joy, but shouted out as loudly as I could,—greatly to the astonishment of old George Johnson, the miller, who happened to be in the field where I had planted my stand!

"Now, though I had obtained what I wanted—a fairly good instrument,—still I was not quite satisfied; as I had produced it by a fortunate chance, and not by skill alone. I therefore set to work again on the other disc of glass, to try if I could finish it in such a way as to excel the first one. After nearly a year's work I found that I could only succeed in equalling it. But then, during this time, I had removed the working of mirrors from mere chance to a fair amount of certainty. By bringing my mathematical knowledge to bear on the subject, I had devised a method of testing and measuring my work which, I am happy to say, has been fairly successful, and has enabled me to produce the spherical, elliptic, parabolic, or hyperbolic curve in my mirrors, with almost unvarying success. The study of the practical working of specula and lenses has also absorbed a good deal of my spare time during the last two years, and the work involved has been scarcely less difficult. Altogether, I consider this last year (1882-3) to mark the busiest period of my life.

"It will be observed that I have only given an account of those branches of study in which I have put to practical test the deductions from theoretical reasoning. I am at present engaged on the theory of the achromatic object-glass, with regard to spherical chromatism—a subject upon which, I believe, nearly all our text-books are silent, but one nevertheless of vital importance to the optician. I can only proceed very slowly with it, on account of having to grind and figure lenses for every step of the theory, to keep myself in the right track; as mere theorizing is apt to lead one very much astray, unless it be checked by constant experiment. For this particular subject, lenses must be ground firstly to spherical, and then to curves of conic sections, so as to eliminate spherical aberration from each lens; so that it will be observed that this subject is not without its difficulties.

"About a month ago (September, 1883), I determined to put to the test the statement of some of our theorists, that the surface of a rotating fluid is either a parabola or a hyperbola. I found by experiment that it is neither, but an approximation to the tractrix (a modification of the catenary), if anything definite; as indeed one, on thinking over the matter, might feel certain it would be—the tractrix being the curve of least friction.

"In astronomy, I have really done very little beyond mere algebraical working of the fundamental theorems, and a little casual observation of the telescope. So far, I must own, I have taken more pleasure in the theory and construction of the telescope, than in its use."

Such is Samuel Lancaster's history of the growth and development of his mind. I do not think there is anything more interesting in the 'Pursuit of Knowledge under Difficulties.' His life has been a gallant endeavour to win further knowledge, though too much at the expense of a constitution originally delicate. He pursues science with patience and determination, and wooes truth with the ardour of a lover. Eulogy of his character would here be unnecessary; but, if he takes due care of his health, we shall hear more of him.[11]

More astronomers in humble life! There seems to to be no end of them. There must be a great fascination in looking up to the heavens, and seeing those wondrous worlds careering in the far-off infinite. Let me look back to the names I have introduced in this chapter of autobiography. First, there was my worthy porter friend at Coupar Angus station, enjoying himself with his three-inch object-glass. Then there was the shoemaker and teacher, and eventually the first-rate maker of achromatic instruments. Look also at the persons whom he supplied with his best telescopes. Among them we find princes, baronets, clergymen, professors, doctors, solicitors, manufacturers, and inventors. Then we come to the portrait painter, who acquired the highest supremacy in the art of telescope making; then to Mr. Lassell, the retired brewer, whose daughters presented his instrument to the nation; and, lastly, to the extraordinary young schoolmaster of Bainbridge, in Yorkshire. And now before I conclude this last chapter, I have to relate perhaps the most extraordinary story of all—that of another astronomer in humble life, in the person of a slate counter at Port Penrhyn, Bangor, North Wales.

While at Birnam, I received a letter from my old friend the Rev. Charles Wicksteed, formerly of Leeds, calling my attention to this case, and inclosing an extract from the letter of a young lady, one of his correspondents at Bangor. In that letter she said: "What you write of Mr. Christmas Evans reminds me very much of a visit I paid a few evenings ago to an old man in Upper Bangor. He works on the Quay, but has a very decided taste for astronomy, his leisure time being spent in its study, with a great part of his earnings. I went there with some friends to see an immense telescope, which he has made almost entirely without aid, preparing the glasses as far as possible himself, and sending them away merely to have their concavity changed. He showed us all his treasures with the greatest delight, explaining in English, but substituting Welsh when at a loss. He has scarcely ever been at school, but has learnt English entirely from books. Among other things he showed us were a Greek Testament and a Hebrew Bible, both of which he can read. His largest telescope, which is several yards long, he has named 'Jumbo,' and through it he told us he saw the snowcap on the pole of Mars. He had another smaller telescope, made by himself, and had a spectroscope in process of making. He is now quite old, but his delight in his studies is still unbounded and unabated. It seems so sad that he has had no right opportunity for developing his talent."

Mr. Wicksteed was very much interested in the case, and called my attention to it, that I might add the story to my repertory of self-helping men. While at York I received a communication from Miss Grace Ellis, the young lady in question, informing me of the name of the astronomer—John Jones, Albert Street, Upper Bangor—and intimating that he would be glad to see me any evening after six. As railways have had the effect of bringing places very close together in point of time—making of Britain, as it were, one great town—and as the autumn was brilliant, and the holiday season not at an end, I had no difficulty in diverging from my journey, and taking Bangor on my way homeward. Starting from York in the morning, and passing through Leeds, Manchester, and Chester, I reached Bangor in the afternoon, and had my first interview with Mr. Jones that very evening.

I found him, as Miss Grace Ellis had described, active, vigorous, and intelligent; his stature short, his face well-formed, his eyes keen and bright. I was first shown into his little parlour downstairs, furnished with his books and some of his instruments; I was then taken to his tiny room upstairs, where he had his big reflecting telescope, by means of which he had seen, through the chamber window, the snowcap of Mars. He is so fond of philology that I found he had no fewer than twenty-six dictionaries, all bought out of his own earnings. "I am fond of all knowledge," he said—"of Reuben, Dan, and Issachar; but I have a favourite, a Benjamin, and that is Astronomy. I would sell all of them into Egypt, but preserve my Benjamin." His story is briefly as follows:—

"I was born at Bryngwyn Bach, Anglesey, in 1818, and I am sixty-five years old. I got the little education I have, when a boy. Owen Owen, who was a cousin of my mother's, kept a school at a chapel in the village of Dwyrain, in Anglesey. It was said of Owen that he never had more than a quarter of a year's schooling, so that he could not teach me much. I went to his school at seven, and remained with him about a year. Then he left; and some time afterwards I went for a short period to an old preacher's school, at Brynsieneyn chapel. There I learnt but little, the teacher being negligent. He allowed the children to play together too much, and he punished them for slight offences, making them obstinate and disheartened. But I remember his once saying to the other children, that I ran through my little lesson 'like a coach.' However, when I was about twelve years old, my father died, and in losing him I lost almost all the little I had learnt during the short periods I had been at school. Then I went to work for the farmers.

"In this state of ignorance I remained for years, until the time came when on Sunday I used to saddle the old black mare for Cadwalladr Williams, the Calvinist Methodist preacher, at Pen Ceint, Anglesey; and after he had ridden away, I used to hide in his library during the sermon, and there I learnt a little that I shall not soon forget. In that way I had many a draught of knowledge, as it were, by stealth. Having a strong taste for music, I was much attracted by choral singing; and on Sundays and in the evenings I tried to copy out airs from different books, and accustomed my hand a little to writing. This tendency was, however, choked within me by too much work with the cattle, and by other farm labour. In a word, I had but little fair weather in my search for knowledge. One thing enticed me from another, to the detriment of my plans; some fair Eve often standing with an apple in hand, tempting me to taste of that.

"The old preacher's books at Pen Ceint were in Welsh. I had not yet learned English, but tried to learn it by comparing one line in the English New Testament with the same line in the Welsh. This was the Hamiltonian method, and the way in which I learnt most languages. I first got an idea of astronomy from reading 'The Solar System,' by Dr. Dick, translated into Welsh by Eleazar Roberts of Liverpool. That book I found on Sundays in the preacher's library; and many a sublime thought it gave me. It was comparatively easy to understand.

"When I was about thirty I was taken very ill, and could no longer work. I then went to Bangor to consult Dr. Humphrys. After I got better I found work at the Port at 12s. a week. I was employed in counting the slates, or loading the ships in the harbour from the railway trucks. I lodged in Fwn Deg, near where Hugh Williams, Gatehouse, then kept a navigation school for young sailors. I learnt navigation, and soon made considerable progress. I also learnt a little arithmetic. At first nearly all the young men were more advanced than myself; but before I left matters were different, and the Scripture words became verified—"the last shall be first." I remained with Hugh Williams six months and a half. During that time I went twice through the

'Tutor's Assistant,' and a month before I left I was taught mensuration. That is all the education I received, and the greater part of it was during my by-hours.

"I got to know English pretty well, though Welsh was the language of those about me. From easy books I went to those more difficult. I was helped in my pronunciation of English by comparing the words with the phonetic alphabet, as published by Thomas Gee of Denbigh, in 1853. With my spare earnings I bought books, especially when my wages began to rise. Mr. Wyatt, the steward, was very kind, and raised my pay from time to time at his pleasure. I suppose I was willing, correct, and faithful. I improved my knowledge by reading books on astronomy. I got, amongst others, 'The Mechanism of the Heavens,' by Denison Olmstead, an American; a very understandable book. Learning English, which was a foreign language to me, led me to learn other languages. I took pleasure in finding out the roots or radixes of words, and from time to time I added foreign dictionaries to my little library. But I took most pleasure in astronomy.

"The perusal of Sir John Herschel's 'Outlines of Astronomy,' and of his 'Treatise on the Telescope,' set my mind on fire. I conceived the idea of making a telescope of my own, for I could not buy one. While reading the Mechanics' Magazine I observed the accounts of men who made telescopes. Why should not I do the same? Of course it was a matter of great difficulty to one who knew comparatively little of the use of tools. But I had a willing mind and willing hands. So I set to work. I think I made my first telescope about twenty years ago. It was thirty-six inches long, and the tube was made of pasteboard. I got the glasses from Liverpool for 4s. 6d. Captain Owens, of the ship Talacra, bought them. He also bought for me, at a bookstall, the Greek Lexicon and the Greek New Testament, for which he paid 7s. 6d. With my new telescope I could see Jupiter's four satellites, the craters on the moon, and some of the double stars. It was a wonderful pleasure to me.

"But I was not satisfied with the instrument. I wanted a bigger and a more perfect one. I sold it and got new glasses from Solomon of London, who was always ready to trust me. I think it was about the year 1868 that I began to make a reflecting telescope. I got a rough disc of glass, from St. Helens, of ten inches diameter. It took me from nine to ten days to grind and polish it ready for parabolising and silvering. I did this by hand labour with the aid of emery, but without a lathe. I finally used rouge instead of emery in grinding down the glass, until I could see my face in the mirror quite plain. I then sent the 8 3/16 inch disc to Mr. George Calver, of Chelmsford, to turn my spherical curve to a parabolic curve, and to silver the mirror, for which I paid him 5L. I mounted this in my timber tube; the focus was ten feet. When everything was complete I tried my instrument on the sky, and found it to have good defining power. The diameter of the other glass I have made is a little under six inches.

"You ask me if their performance satisfies me? Well; I have compared my six-inch reflector with a 4 1/4 inch refractor, through my window, with a power of 100 and 140. I can't say which was the best. But if out on a clear night I think my reflector would take more power than the refractor. However that may be, I saw the snowcap on the planet Mars quite plain; and it is satisfactory to me so far. With respect to the 8 3/16 inch glass, I am not quite satisfied with it yet; but I am making improvements, and I believe it will reward my labour in the end."

Besides these instruments John Jones has an equatorial which is mounted on a tripod stand, made by himself. It contains the right ascension, declination, and azimuth index, all neatly carved upon slate. In his spectroscope he makes his prisms out of the skylights used in vessels. These he grinds down to suit his purpose. I have not been able to go into the complete detail of the manner in which he effects the grinding of his glasses. It is perhaps too technical to be illustrated in words, which are full of focuses, parabolas, and convexities. But enough may be gathered from the above account to give an idea of the wonderful tenacity of this aged student, who counts his slates into the ships by day, and devotes his evenings to the perfecting of his astronomical instruments. But not only is he an astronomer and a philologist; he is also a bard, and his poetry is much admired in the district. He writes in Welsh, not in English, and signs himself "Ioan, of Bryngwyn Bach," the place where he was born. Indeed, he is still at a loss for words when he speaks in English. He usually interlards his conversation with passages in Welsh, which is his mother-tongue. A friend has, however, done me the favour to translate two of John Jones's poems into English. The first is 'The Telescope':—

"To Heaven it points, where rules the Sun
In golden gall'ries bright;
And the pale Moon in silver rays
Makes dalliance in the night.

"It sweeps with eagle glances
The sky, its myriad throng,
That myriad throng to marshal
And bring to us their song.

"Orb upon orb it follows
As oft they intertwine,
And worlds in vast processions
As if in battle line.

"It loves all things created,
To follow and to trace;
And never fears to penetrate
The dark abyss of space."

The next is to 'The Comet':—

"A maiden fair, with light of stars bedecked,
Starts out of space at Jove's command;
With visage wild, and long dishevelled hair,
Speeds she along her starry course;
The hosts of heaven regards she not,—
Fain would she scorn them all except her father Sol,
Whose mighty influence her headlong course doth all control."

The following translation may also be given: it shows that the bard is not without a spice of wit. A fellow-workman teased him to write some lines; when John Jones, in a seemingly innocent manner, put some questions, and ascertained that he had once been a tailor. Accordingly this epigram was written, and appeared in the local paper the week after: "To a quondam Tailor, now a Slate-teller":—

"To thread and needle now good-bye,
With slates I aim at riches;
The scissors will I ne'er more ply,
Nor make, but order, breeches."[12]

The bi-lingual speech is the great educational difficulty of Wales. To get an entrance into literature and science requires a knowledge of English; or, if not of English, then of French or German. But the Welsh language stands in the way. Few literary or scientific works are translated into Welsh. Hence the great educational difficulty continues, and is maintained from year to year by patriotism and Eisteddfods.

Possibly the difficulties to be encountered may occasionally evoke unusual powers of study; but this can only occur in exceptional cases. While at Bangor Mr. Cadwalladr Davies read to me the letter of a student and professor, whose passion for knowledge is of an extraordinary character. While examined before the Parliamentary Committee appointed to inquire into the condition of intermediate and higher education in Wales and Monmouthshire, Mr. Davies gave evidence relating to this and other remarkable cases, of which the following is an abstract, condensed by himself:—

"The night schools in the quarry districts have been doing a very great work; and, if the Committee will allow me, I will read an extract from a letter which I received from Mr. Bradley Jones, master of the Board Schools at Llanarmon, near Mold, Flintshire, who some years ago kept a very flourishing night school in the neighbourhood. He says: 'During the whole of the time (fourteen years) that I was at Carneddi, I carried on these schools, and I believe I have had more experience of such institutions than any teacher in North Wales. For several years about 120 scholars used to attend the Carneddi night school in the winter months, four evenings a week. Nearly all were quarrymen, from fourteen to twenty-one years of age, and engaged at work from 7 A.M. to 5.30 P.M. So intense was their desire for education that some of them had to walk a distance of two or even three miles to school. These, besides working hard all day, had to walk six miles in the one case and nine in the other before school-time, in addition to the walk home afterwards. Several of them used to attend all the year round, even coming to me for lessons in summer before going to work, as well as in the evening. Indeed, so anxious were some of them, that they would often come for lessons as early as five o'clock in the morning. This may appear almost incredible, but any of the managers of the Carneddi School could corroborate the statement.'

"I have now in my mind's eye," continues Mr. Bradley, "several of these young men, who, by dint of indefatigable labour and self-denial, ultimately qualified themselves for posts in which a good education is a sine qua non. Some of them are to-day quarry managers, professional men, certificated teachers, and ministers of the Gospel. Five of them are at the present time students at Bala College. One got a situation in the Glasgow Post Office as letter-carrier. During his leisure hours he attended the lectures at one of the medical schools of that city, and in course of time gained his diploma. He is now practising as a surgeon, and I understand with signal success. This gentleman worked in the Penrhyn Quarry until he was twenty years old. I could give many more instances of the resolute and self-denying spirit with which the young quarrymen of Bethesda sought to educate themselves. The teachers of the other schools in that neighbourhood could give similar examples, for during the winter months there used to be no less than 300 evening scholars under instruction in the different schools. The Bethesda booksellers could tell a tale that would surprise our English friends. I have been informed by one of them that he has sold to young quarrymen an immense number of such works as Lord Macaulay's, Stuart Mill's, and Professor Fawcett's; and it is no uncommon sight to find these and similar works read and studied by the young quarrymen during the dinner hour."

"I can give," proceeds Mr. Cadwalladr Davies, "one remarkable instance to show the struggles which young Welshmen have to undertake in order to get education. The boy in question, the son of 'poor but honest parents,' left the small national school of his native village when he was 12 1/2 years of age, and then followed his father's occupation of shoemaking until he was 16 1/2 years of age. After working hard at his trade for four years, he, his brother, and two fellow apprentices, formed themselves into a sort of club to learn shorthand, the whole matter being kept a profound secret. They had no teachers, and they met at the gas-works, sitting opposite the retorts on a bench supported at each end with bricks. They did not penetrate far into the mysteries of Welsh shorthand; they soon abandoned the attempt, and induced the village schoolmaster to open a night school.

"This, however, did not last long. The young Crispin was returning late one night from Llanrwst in company with a lad of the same age, and both having heard much of the blessings of education from a Scotch lady who took a kindly interest in them, their ambition was inflamed, and they entered into a solemn compact that they would thenceforward devote themselves body and soul to the attainment of an academical degree. Yet they were both poor. One was but a shoemaker's apprentice, while the other was a pupil teacher earning but a miserable weekly pittance. One could do the parts of speech; the other could not. One had struggled with the pans asinorum; the other had never seen it. I may mention that the young pupil teacher is now a curate in the Church of England. He is a graduate of Cambridge University and a prizeman of Clare College. But to return to the little shoemaker.

"After returning home from Llanrwst, he disburthened his heart to his mother, and told her that shoemaking, which until now he had pursued with extraordinary zest, could no longer interest him. His mother, who was equal to the emergency, sent the boy to a teacher of the old school, who had himself worked his way from the plough. After the exercise of considerable diplomacy, an arrangement was arrived at whereby the youth was to go to school on Mondays, Wednesdays, and Fridays, and make shoes during the remaining days of the week. This suited him admirably. That very night he seized upon a geography, and began to learn the counties of England and Wales. The fear of failure never left him for two hours together, except when he slept. The plan of work was faithfully kept; though by this time shoemaking had lost its charms. He shortened his sleeping hours, and rose at any moment that he awoke—at two, three, or four in the morning. He got his brother, who had been plodding with him over shorthand, to study horticulture, and fruit and vegetable culture; and that brother shortly after took a high place in an examination held by the Royal Horticultural Society. For a time, however, they worked together; and often did their mother get up at four o'clock in the depth of winter, light their fire, and return to bed after calling them up to the work of self-culture. Even this did not satisfy their devouring ambition. There was a bed in the workshop, and they obtained permission to sleep there. Then they followed their own plans. The young gardener would sit up till one or two in the morning, and wake his brother, who had gone to bed as soon as he had given up work the night before.

Now he got up and studied through the small hours of the morning until the time came when he had to transfer his industry to shoemaking, or go to school on the appointed days after the distant eight o'clock had come. His brother had got worn out. Early sleep seemed to be the best. They then both went to bed about eight o'clock, and got the policeman to call them up before retiring himself.

"So the struggle went on, until the faithful old schoolmaster thought that his young pupil might try the examination at the Bangor Normal College. He was now eighteen years of age; and it was eighteen months since the time when he began to learn the counties of England and Wales. He went to Bangor, rigged out in his brother's coat and waistcoat, which were better than his own; and with his brother's watch in his pocket to time himself in his examinations. He went through his examination, but returned home thinking he had failed. Nevertheless, he had in the meantime, on the strength of a certificate which he had obtained six months before, in an examination held by the Society of Arts and Sciences in Liverpool, applied for a situation as teacher in a grammar-school at Ormskirk in Lancashire. He succeeded in his application, and had been there for only eight days when he received a letter from Mr. Rowlands, Principal of the Bangor Normal College, informing him that he had passed at the head of the list, and was the highest non-pupil teacher examined by the British and Foreign Society. Having obtained permission from his master to leave, he packed his clothes and his few books. He had not enough money to carry him home; but, unasked, the master of the school gave him 10s. He arrived home about three o'clock on a Sunday morning, after a walk of eleven miles over a lonely road from the place where the train had stopped. He reeled on the way, and found the country reeling too. He had been sleeping eight nights in a damp bed. Six weeks of the Bangor Session passed, and during that time he had been delirious, and was too weak to sit up in bed. But the second time he crossed the threshold of his home he made for Bangor and got back his "position," which was all important to him, and he kept it all through.

"Having finished his course at Bangor he went to keep a school at Brynaman; he endeavoured to study but could not. After two years he gave up the school, and with 60L. saved he faced the world once more. There was a scholarship of the value of 40L. a year, for three years, attached to one of the Scotch Universities, to be competed for. He knew the Latin Grammar, and had, with help, translated one of the books of Caesar. Of Greek he knew nothing, save the letters and the first declension of nouns; but in May he began to read in earnest at a farmhouse. He worked every day from 6 A.M. to 12 P.M. with only an hour's intermission. He studied the six Latin and two Greek books prescribed; he did some Latin composition unaided; brushed up his mathematics; and learnt something of the history of Greece and Rome. In October, after five months of hard work, he underwent an examination for the scholarship, and obtained it; beating his opponent by twenty-eight marks in a thousand. He then went up to the Scotch University and passed all the examinations for his ordinary M.A. degree in two years and a half. On his first arrival at the University he found that he could not sleep; but he wearily yet victoriously plodded on; took a prize in Greek, then the first prize in philosophy, the second prize in logic, the medal in English literature, and a few other prizes.

"He had 40L. when he first arrived in Scotland; and he carried away with him a similar sum to Germany, whither he went to study for honours in philosophy. He returned home with little in his pocket, borrowing money to go to Scotland, where he sat for honours and for the scholarship. He got his first honours, and what was more important at the time, money to go on with. He now lives on the scholarship which he took at that time; is an assistant professor; and, in a fortnight, will begin a course of lectures for ladies in connection with his university. Writing to me a few days ago,[13] he says, 'My health, broken down with my last struggle, is quite restored, and I live with the hope of working on. Many have worked more constantly, but few have worked more intensely. I found kindness on every hand always, but had I failed in a single instance I should have met with entire bankruptcy. The failure would have been ruinous.... I thank God for the struggle, but would not like to see a dog try it again. There are droves of lads in Wales that would creep up but they cannot. Poverty has too heavy a hand for them.'"

The gentleman whose brief history is thus summarily given by Mr. Davies, is now well known as a professor of philosophy; and, if his health be spared, he will become still better known. He is the author of several important works on 'Moral Philosophy,' published by a leading London firm; and more works are announced from his pen. The victorious struggle for knowledge which we have recounted might possibly be equalled, but it could not possibly be surpassed. There are, however, as Mr. Davies related to the Parliamentary Committee, many instances of Welsh students—most of them originally quarrymen—who keep themselves at school by means of the savings effected from manual labour, "in frequent cases eked out and helped by the kindness of friends and neighbours," who struggle up through many difficulties, and eventually achieve success in the best sense of the term. "One young man"—as the teacher of a grammar-school, within two miles of Bangor, related to Mr. Davies—"who came to me from the quarry some time ago, was a gold medallist at Edinburgh last winter;" and contributions are readily made by the quarrymen to help forward any young man who displays an earnest desire for knowledge in science and literature.

It is a remarkable fact that the quarrymen of Carnarvonshire have voluntarily contributed large sums of money towards the establishment of the University College in North Wales—the quarry districts in that county having contributed to that fund, in the course of three years, mostly in half-crown subscriptions, not less than 508L. 4s. 4d.—"a fact," says Mr. Davies, "without its parallel in the history of the education of any country;" the most striking feature being, that these collections were made in support of an institution from which the quarrymen could only very remotely derive any benefit.

While I was at Bangor, on the 24th of August, 1883, the news arrived that the Committee of Selection had determined that Bangor should be the site for the intended North Wales University College. The news rapidly spread, and great rejoicings prevailed throughout the borough, which had just been incorporated. The volunteer band played through the streets; the church bells rang merry peals; and gay flags were displayed from nearly every window. There never was such a triumphant display before in the cause of University education.

As Mr. Cadwalladr Davies observed at the banquet, which took place on the following day: "The establishment of the new institution will mark the dawn of a new era in the history of the Welsh people. He looked to it, not only as a means of imparting academical knowledge to the students within its walls, but also as a means of raising the intellectual and moral tone of the whole people. They were fond of quoting the saying of a great English writer, that there was something Grecian in the Celtic race, and that the Celtic was the refining element in the British character; but such remarks, often accompanied as they were with offensive comparisons from Eisteddfod platforms, would in future be put to the test, for they would, with their new educational machinery, be placed on a footing of perfect equality with the Scotch and the Irish people."

And here must come to an end the character history of my autumn tour in Ireland, Scotland, Yorkshire, and Wales. I had not the remotest intention when setting out of collecting information and writing down my recollections of the journey. But the persons I met, and the information I received, were of no small interest—at least to myself; and I trust that the reader will derive as much pleasure from perusing my observations as I have had in collecting and writing them down. I do think that the remarkable persons whose history and characters I have endeavoured, however briefly, to sketch, will be found to afford many valuable and important lessons of Self-Help; and to illustrate how the moral and industrial foundations of a country may be built up and established.

Footnotes for Chapter XII.

[1] A "poet," who dates from "New York, March 1883," has published seven stanzas, entitled "Change here for Blairgowrie," from which we take the following:—

"From early morn till late at e'en,
John's honest face is to be seen,
Bustling about the trains between,
Be 't sunshine or be 't showery;
And as each one stops at his door,
He greets it with the well-known roar
Of 'Change here for Blairgowrie.'
Even when the still and drowsy night
Has drawn the curtains of our sight,
John's watchful eyes become more bright,
And take another glow'r aye
Thro' yon blue dome of sparkling stars
Where Venus bright and ruddy Mars
Shine down upon Blairgowrie.
He kens each jinkin' comet's track,
And when it's likely to come back,
When they have tails, and when they lack—
In heaven the waggish power aye;
When Jupiter's belt buckle hings,
And the Pyx mark on Saturn's rings,
He sees from near Blairgowrie."

[2] The Observatory, No. 61, p. 146; and No. 68, p. 371.

[3] In an article on the subject in the Dundee Evening Telegraph, Mr. Robertson observes: "If our finite minds were more capable of comprehension, what a glorious view of the grandeur of the Deity would be displayed to us in the contemplation of the centre and source of light and heat to the solar system. The force requisite to pour such continuous floods to the remotest parts of the system must ever baffle the mind of man to grasp. But we are not to sit down in indolence: our duty is to inquire into Nature's works, though we can never exhaust the field. Our minds cannot imagine motion without some Power moving through the medium of some subordinate agency, ever acting on the sun, to send such floods of light and heat to our otherwise cold and dark terrestrial ball; but it is the overwhelming magnitude of such power that we are incapable of comprehending. The agency necessary to throw out the floods of flame seen during the few moments of a total eclipse of the sun, and the power requisite to burst open a cavity in its surface, such as could entirely engulph our earth, will ever set all the thinking capacity of man at nought."

[4] The Observatory, Nos. 34, 42, 45, 49, and 58.

[5] We regret to say that Sheriff Barclay died a few months ago, greatly respected by all who knew him.

[6] Sir E. Denison Beckett, in his Rudimentary Treatise on clocks and Watches and Bells, has given an instance or the telescope-driving clock, invented by Mr. Cooke (p. 213).

[7] J. Norman Lockyer, F.R.S.—Stargazing, Past and Present, p. 302.

[8] This excellent instrument is now in the possession of my son-in-law, Dr. Hartree, of Leigh, near Tunbridge.

[9] An interesting account of Mr. Alvan Clark is given in Professor Newcomb's 'Popular Astronomy,' p. 137.

[10] A photographic representation of this remarkable telescope is given as the frontispiece to Mr. Lockyer's Stargazing, Past and Present; and a full description of the instrument is given in the text of the same work. This refracting telescope did not long remain the largest. Mr. Alvan Clark was commissioned to erect a larger equatorial for Washington Observatory; the object-glass (the rough disks of which were also furnished by Messrs. Chance of Birmingham) exceeding in aperture that of Mr. Cooke's by only one inch. This was finished and mounted in November, 1873. Another instrument of similar size and power was manufactured by Mr. Clark for the University of Virginia. But these instruments did not long maintain their supremacy. In 1881, Mr. Howard Grubb, of Dublin, manufactured a still larger instrument for the Austrian Government—the object-glass being of twenty-seven inches aperture. But Mr. Alvan Clark was not to be beaten. In 1882, he supplied the Russian Government with the largest refracting telescope in existence the object-glass being of thirty inches diameter. Even this, however, is to be surpassed by the lens which Mr. Clark has in hand for the Lick Observatory (California), which is to have a clear aperture of three feet in diameter.

[11] Since the above passage was written and in type, I have seen (in September 1884) the reflecting telescope referred to at pp. 357-8. It was mounted on its cast-iron equatorial stand, and at work in the field adjoining the village green at Bainbridge, Yorkshire. The mirror of the telescope is 8 inches in diameter; its focal length, 5 feet;

and the tube in which it is mounted, about 6 feet long. The instrument seemed to me to have an excellent defining power.

But Mr. Lancaster, like every eager astronomer, is anxious for further improvements. He considers the achromatic telescope the king of instruments, and is now engaged in testing convex optical surfaces, with a view to achieving a telescope of that description. The chief difficulty is the heavy charge for the circular blocks of flint glass requisite for the work which he meditates. "That," he says, "is the great difficulty with amateurs of my class." He has, however, already contrived and constructed a machine for grinding and polishing the lenses in an accurate convex form, and it works quite satisfactorily. Mr. Lancaster makes his own tools. From the raw material, whether of glass or steel, he produces the work required. As to tools, all that he requires is a bar of steel and fire; his fertile brain and busy hands do the rest. I looked into the little workshop behind his sitting-room, and found it full of ingenious adaptations. The turning lathe occupies a considerable part of it; but when he requires more space, the village smith with his stithy, and the miller with his water-power, are always ready to help him. His tools, though not showy, are effective. His best lenses are made by himself: those which he buys are not to be depended upon. The best flint glass is obtained from Paris in blocks, which he divides, grinds, and polishes to perfect form.

I was attracted by a newly made machine, placed on a table in the sitting-room; and on inquiry found that its object was to grind and polish lenses. Mr. Lancaster explained that the difficulty to be overcome in a good machine, is to make the emery cut the surface equally from centre to edge of the lens, so that the lens will neither lengthen nor shorten the curve during its production. To quote his words: "This really involves the problem of the 'three bodies,' or disturbing forces so celebrated in dynamical mathematics, and it is further complicated by another quantity, the 'coefficient of attrition,' or work done by the grinding material, as well as the mischief done by capillary attraction and nodal points of superimposed curves in the path of the tool. These complications tend to cause rings or waves of unequal wear in the surface of the glass, and ruin the defining power of the lens, which depends upon the uniformity of its curve. As the outcome of much practical experiment, combined with mathematical research, I settled upon the ratio of speed between the sheave of the lens-tool guide and the turn-table; between whose limits the practical equalization of wear (or cut of the emery) might with the greater facility be adjusted, by means of varying the stroke and eccentricity of the tool. As the result of these considerations in the construction of the machine, the surface of the glass 'comes up' regularly all over the lens; and the polishing only takes a few minutes' work—thus keeping the truth of surface gained by using a rigid tool."

The machine in question consists of a revolving sheave or ring, with a sliding strip across its diameter; the said strip having a slot and clamping screw at one end, and a hole towards the other, through which passes the axis of the tool used in forming the lens,—the slot in the strip allowing the tool to give any stroke from 0 to 1.25 inch. The lens is carried on a revolving turn-table, with an arrangement to allow the axis of the lens to coincide with the axis of the table. The ratio of speed between the sheave and turn-table is arranged by belt and properly sized pulleys, and the whole can be driven either by hand or by power. The sheave merely serves as a guide to the tool in its path, and the lens may either be worked on the turn-table or upon a chuck attached to the tool rod. The work upon the lens is thus to a great extent independent of the error of the machine through shaking, or bad fitting, or wear; and the only part of the machine which requires really first-class work is the axis of the turn-table, which (in this machine) is a conical bearing at top, with steel centre below,—the bearing turned, hardened, and then ground up true, and run in anti-friction metal. Other details might be given, but these are probably enough for present purposes. We hope, at some future time, for a special detail of Mr. Lancaster's interesting investigations, from his own mind and pen.

[12] The translations are made by W. Cadwalladr Davies, Esq.

[13] This evidence was given by Mr. W. Cadwalladr Davies on the 28th October, 1880.

Printed by BoD™ in Norderstedt, Germany

9 781803 622484